Frederic Ward Putnam, Institute Essex

## The Naturalists' Directory.

Part II. North America and the West Indies

Frederic Ward Putnam, Institute Essex

**The Naturalists' Directory.**
*Part II. North America and the West Indies*

ISBN/EAN: 9783337025144

Printed in Europe, USA, Canada, Australia, Japan

Cover: Foto ©berggeist007 / pixelio.de

More available books at **www.hansebooks.com**

THE

# NATURALISTS' DIRECTORY.

---

## PART II.

---

### NORTH AMERICA AND THE WEST INDIES.

SALEM:

PUBLISHED BY THE ESSEX INSTITUTE.

1866.

# INTRODUCTION.

———o———

In this part of the "Naturalists' Directory" the addresses of the various persons are given under a systematical arrangement. A geographical grouping of the names and an alphabetical index will follow. The number preceding each name is given to facilitate indexing. This part will be issued with the "Proceedings," several pages at a time, as fast as it can be prepared.

The Editor returns his thanks to the numerous friends who have aided him, since the publication of the first part, by sending new names and corrections, and earnestly hopes that his attention may be called to any mistakes or omissions noticed in this second part, at as early a day as possible.

Information respecting Naturalists residing in the Southern States and Mexico is especially requested.

Those persons who have not answered the circulars forwarded to them are requested to do so, even if their addresses are correctly given, as it is only from the personal statement of each that perfect confidence can be had in the printed list. In regard to this subject, the Editor would state that answers have been received to the greater part of the circulars issued. The addresses which are left doubtful from the non receipt of answers, will be designated at the close of this part.

Notices of the decease or change of address, of persons whose names appear in the Directory are particularly requested.

Owing to the large number of names which have been received since the first twenty pages of this part of the Directory were issued, and the various corrections which have been made, it has been thought best to reprint them, especially as it was desirable to slightly change the arrangement of the names. Subscribers to the work will therefore please substitute this edition for the former.

F. W. PUTNAM,
*Editor.*

ESSEX INSTITUTE, SALEM, MASS.,
May 1, 1866.

# NATURALISTS' DIRECTORY.

## GEOLOGY.

1. Prof. LOUIS AGASSIZ (Professor of Zoölogy and Geology, Harvard University; Director and Curator, Museum of Comparative Zoölogy), Cambridge, Mass. *General.*
2. Prof. WM. E. A. AIKIN (Professor of Chemistry and Pharmacy, University of Maryland; Swan Lecturer on applied Chemistry, Maryland Institute), 25 Hamilton street, Baltimore, Md. *North American.*
3. HENRY D'ALIGNY (Mining Engineer; Resident Agent, St. Mary's Canal Mineral Land Co.), Houghton, L. S., Mich. *North American.*
4. Prof. E. B. ANDREWS (Professor of Chemistry, Mineralogy and Geology, Marietta College), Marietta, Ohio. *North American.*
5. AUSTIN BACON, Natick, Mass. *Local.*
6. Prof. L. W. BAILEY (Professor of Chemistry and Natural History, University of New Brunswick), Fredericton, New Brunswick. *Local.*
7. D. M. BALCH (Chemist), Salem, Mass. *Local.*
8. Rev. M. W. BEAUCHAMP, King's Ferry, Cayuga, N. Y. *Local.*
9. GEORGE BECK, Lockport, N. Y. *Local.*
10. Prof. ROBERT BELL (Professor of Natural History, Chemistry and Geology, Queen's University; Assistant, Geological Survey of Canada; Secretary, Botanical Society of Canada), Kingston, Canada West. *North American.*
11. Prof. JAMES G. BLAIR, Athens, Ohio. *North American.*
12. Prof. W. P. BLAKE (Director and Professor of Mineralogy, Geology and Mining, Mining and Agricultural College), Post Office box 2077, San Francisco, Cal. *North American.*
13. FRANK H. BRADLEY (Curator of Geology, Yale College), New Haven, Ct. *North American.*
14. G. C. BROADHEAD, Pleasant Hill, Cass Co., Mo. *North American.*
15. Prof. EZRA L. CARR (Professor of Chemistry and Natural History, Wisconsin State University), Beloit, Wis. *Local.*
16. SAMUEL R. CARTER, Paris Hill, Oxford Co., Me. *Local.*

17. Prof. CHARLES F. CHANDLER (Professor of Chemistry, School of Mines, Columbia College), East Forty-ninth street, New, York, N. Y. *North American.*

18. Prof. EDWARD J. CHAPMAN (Professor of Mineralogy and Geology, University College), Toronto, Canada West. *North American.*

19. T. APOLEON CHENEY (Librarian, Georgic Library), Havana, N. Y. *Local.*

20. CHANDLER CHILDS, Desmoines, Iowa. *Local.*

21. Prof. GEORGE H. COOK (State Geologist of New Jersey; Professor of Chemistry and Natural History, Rutgers College), New Brunswick, N. J. *North American.*

22. Rev. SYLVESTER COWLES, West Randolph Cattaraugus Co., N. Y. *Local.*

23. E. T. COX, New Harmony, Ind. *Local.*

24. Dr. E. S. CROSIER, New Albany, Ind. *Local.*

25. HIRAM A. CUTTING, Lunenburg, Essex Co., Vt. *Local.*

26. Prof. JAMES D. DANA (Professor of Geology and Mineralogy, Yale College), New Haven, Ct. *General.*

27. Prof. J. W. DAWSON (Principal, McGill University), Montreal, Canada. *British North America.*

28. ANDREW DICKSON, Kingston, Canada West. *North American.*

29. H. DODGE, Skaneateles, N. Y. *Local.*

30. Rev. E. B. EDDY, Waltham, Mass. *Local.*

31. Dr. M. N. ELROD, Jeffersonville, Ind. *Local.*

32. L. ENGELBROCHT, Portsmouth, Ohio. *Local.*

33. HENRY ENGELMANN (Mining Engineer), Belleville, Ill. *North American.*

34. Prof. JACOB ENNIS, Philadelphia, Pa. *Local.*

35. C. F. ESCHWEILER (Mining Engineer), No. 110, Springfield street, Boston, Mass. *North American.*

36. Prof. E. W. EVANS, Marietta, Ohio. *North American.*

37. Hon. SAMUEL EWING, Randolph, N. Y. *Local.*

38. Dr. P. J. FARNSWORTH, Lyons, Clinton Co., Iowa. *Local.*

39. J. W. FOSTER (Mining Engineer), Chicago, Ill. *North American.*

40. Hon. GEORGE GEDDES, Fairmount, Onondaga Co., N. Y. *Local.*

41. WM. GOSSIP (Secretary, Nova Scotian Institute of Natural Science), Halifax, Nova Scotia. *Local.*

42. A. D. HAGER (State Geologist of Vermont; Curator of Vermont State Cabinet), Proctorsville, Vt. *North American.*

43. Prof. JAMES HALL (State Geologist of New York, Iowa and Wisconsin), Albany, N. Y. *General.*

44. Rev. SAMUEL R. HALL, Brownington, Vt. *Local.*

45. ISAAC N. HARMON, Chicago, Ill. *Local.*
46. LOUIS HARPER (Mining Engineer), No. 1, Rector street, New York, N. Y. *North American.*
47. F. HAWN, Leavenworth City, Kansas. *North American.*
48. Dr. F. V. HAYDEN, Smithsonian Institution, Washington, D. C. *North American.*
49. E. W. HILGARD (State Geologist of Mississippi), Oxford, Miss. *North American.*
50. S. W. HILL, Houghton, Mich. *Local.*
51. C. H. HITCHCOCK, No. 37, Park Row, New York, N. Y. *North American.*
52. JAMES T. HODGE, Newburg, N. Y. *North American.*
53. Prof. FRANCIS S. HOLMES, College of Charleston, Charleston, S. C. *Local.*
54. Rev. Dr. HONEYMAN, Antigonish, Nova Scotia. *Local.*
55. Prof. EDMUND O. HOVEY (Professor of Chemistry and Geology, Wabash College), Crawfordsville, Ind. *Local.*
56. Prof. HENRY HOW (Professor of Chemistry and Natural History, University of King's College), Windsor, Nova Scotia. *North American.*
57. ROBERT HOWELL, Nichols, Tioga Co., N. Y. *Local.*
58. Prof. O. P. HUBBARD (Professor of Chemistry, Mineralogy and Geology, Dartmouth College), Hanover, N. H. *North American.*
59. Dr. C. T. JACKSON (Vice President, Boston Society of Natural History), Boston, Mass. *North American.*
60. JOHN JENKINS, Monroe, Orange Co., N. Y. *Local.*
61. Prof. JOHN JOHNSTON (Professor of Natural Sciences, Wesleyan University), Middletown, Ct. *North American.*
62. JAMES P. KIMBALL (Mining Engineer), No. 33, Wall street, and 40 St. Mark's Place, New York, N. Y. *North American.*
63. CLARENCE KING (Assistant, California Geological Survey), Irvington, N. Y. and San Francisco, Cal. *North American.*
64. JOHN H. KLIPPART, Columbus, Ohio. *Local.*
65. I. A. LAPHAM, Milwaukee, Wis. *Local.*
66. Dr. GEORGE A. LATHROP, East Saginaw, Mich. *Local.*
67. ISAAC LEA (Vice President, American Philosophical Society), No. 1622, Locust street, Philadelphia, Pa. *North American.*
68. JOSEPH LESLEY, Office Pennsylvania R. R., Philadelphia, Pa. *North American.*
69. J. P. LESLEY (Professor of Mining, University of Pennsylvania; Fourth Secretary and Librarian, American Philosophical Society), No. 1016, Clinton street, Philadelphia, Pa. *North American.*

70. Prof. LEO LESQUEREUX, Columbus, Ohio. *North American.*
71. ELIAS LEWIS (Chairman of the Committee on Natural History, Long Island Historical Society), No. 16, Court street, Brooklyn, N. Y. *Local.*
72. Rev. SAMUEL LOCKWOOD, Keyport, N. J. *Cretaceous. Local.*
73. Sir WILLIAM E. LOGAN (Director, Geological Survey of Canada), Montreal, Canada. *General.*
74. Rev. J. E. LONG, Hublersburg, Centre Co., Pa. *Local.*
75. BENJ. SMITH LYMAN (Mining Engineer), No. 35, South Fifth street, Philadelphia, Pa. *North American.*
76. SYDNEY S. LYON, Jeffersonville, Ind. *North American.*
77. Prof. OLIVER MARCY (Professor of Natural History, North-western University), Evanston, Ill. *North American.*
78. Dr. R. P. MASON, Milford, Ohio. *Local.*
79. G. F. MATTHEW (Curator, Natural History Society of St. John), St. John, New Brunswick. *Local.*
80. Prof. J. H. McCHESNEY, Jacksonville, Ill. *North American.*
81. R. McFARLANE, Fort Anderson, British America. *Local.*
82. F. B. MEEK, Smithsonian Institution, Washington, D. C. *North American.*
83. Hon. ANSON S. MILLER, Rockford, Ill. *Local.*
84. Rev. JAMES E. MILLS, Amherst, Nova Scotia. *North American.*
85. Prof. JOSEPH MOORE (Professor of Natural History, Earlham College), Richmond, Ind. *Local.*
86. BENJAMIN F. MUDGE (Professor of Natural History, Kansas State Agricultural College), Manhatten, Kansas. *North American.*
87. ALEXANDER MURRAY, St. Johns, New Foundland. *North American.*
88. Prof. HENRY B. NASON, Beloit, Wis. *North American.*
89. J. V. C. NELLIS, Auburn, N. Y. *Local.*
90. Dr. J. S. NEWBERRY, Cleveland, Ohio. *North American.*
91. JOHN A. NICHOLS, Poultney, Vt. *Local.*
92. W. H. NILES (Lecturer Massachusetts State Teacher's Institute), New Haven, Ct. *North American.*
93. Prof. J. G. NORWOOD (Professor of Natural Science and Natural Philosophy, Missouri State University), Columbia, Boone County, Mo. *North American.*
94. J. KELLY O'NEALE, Lebanon, Ohio. *Local.*
95. Prof. RICHARD OWEN (Professor of Natural Sciences, Indiana State University), Bloomington, Ind., from Sept. 1 to July 1; rest of the year, New Harmony, Ind. *North American.*
96. Prof. THEODORE S. PARVIN (Professor of Natural History, Iowa State University), Iowa City, Iowa. *North American.*

97. Prof. E. J. Pickett, Rochester, N. Y. *North American.*
98. Dr. William Prescott, Concord, N. H. *Local.*
99. Raphael Pumpelly, Owego, N. Y. *Arizona, China, Japan.*
100. Hon. II. S. Randall, Cortland Village, N. Y. *Local.*
101. Dr. Samuel Reid, New Albany, Ind. *Local.*
102. E. J. Rice, Muncie, Ind. *Local.*
103. James Richards, Litchfield, Ct. *Local.*
104. R. A. Rideout, Garland, Me. *Local.*
105. Prof. Wm. B. Rogers (President, Massachusetts Institute of Technology; Corresponding Secretary, American Academy of Arts and Sciences), Boston, Mass. *General.*
106. Joseph M. Rowell, Lynn, Mass. *Local.*
107. Prof. James M. Safford, (State Geologist of Tennessee), Post Office box 36, Nashville, Tenn. *North American.*
108. Peter W. Sheafer, Pottsville, Schuylkill Co., Pa. *Coal.*
109. Dr. B. F. Shumard, St. Louis, Mo. *North American.*
110. Prof. Benjamin Silliman (Professor of Chemistry, Yale College), New Haven, Ct. *North American.*
111. Dr. R. P. Stevens, New York, N. Y. *North American.*
112. O. H. St. John, Waterloo, Iowa. *Local.*
113. Charles Stodder, No. 75, Kilby street, Boston, Mass. *Local.*
114. R. H. Stretch, Virginia City, Nevada. *North American.*
115. Joseph Sullivant, Columbus, Ohio. *Local.*
116. Prof. G. C. Swallow (State Geologist of Kansas), Columbia, Boone Co., Mo. *North American.*
117. Prof. Sanborn Tenney (Professor of Natural History, Vassar Female College, Poughkeepsie, N. Y. *North American.*
118. W. H. B. Thomas, Mount Holly, N. J. *North American.*
119. C. B. Trego, No. 612, North Thirteenth street, Philadelphia, Pa. *North American.*
120. Dr. John B. Trask, San Francisco, Cal. *Local.*
121. Philip T. Tyson (State Geologist of Maryland), Baltimore, Md. *North American.*
122. Prof. A. E. Verrill, (Mining Engineer; Professor of Zoölogy, Yale College), New Haven, Ct. *North American.*
123. T. C. Walbridge, Belleville, Upper Canada. *Local.*
124. R. P. Whitfield, Albany, N. Y. *North American.*
125. Dr. Charles A. White, Iowa City, Iowa. *North American.*
126. Prof. J. D. Whitney (State Geologist of California), San Francisco, Cal., and Northampton, Mass. *North American.*
127. Col. Charles Whittlesey, Cleveland, Ohio. *North American.*

128. CHARLES P. WILLIAMS, No. 138, Walnut street, Philadelphia, Pa. *North American.*
129. Prof. ALEXANDER WINCHELL (Professor of Natural History, Michigan State University), Ann Arbor, Mich. *North American.*
130. A. H. WORTHEN (State Geologist of Illinois) Springfield, Ill. *North American.*

## MINERALOGY.

131. HENRY D'ALIGNY (Mining Engineer; Resident Agent, St. Mary's Canal Mineral Land Co.), Houghton, L. S., Mich.
132. OSCAR D. ALLEN, Camden, N. J.
133. S. C. H. BAILEY, No. 5, Beekman street, New York, N. Y.
134. EDWARD E. BARDEN, Rockport, Mass.
135. VINCENT BARNARD, Kennett Square, Chester Co., Pa. *A Collector.*
136. WM. BARNES (Mining Engineer), Post Office box 274, Halifax, Nova Scotia.
137. Rev. E. R. BEADLE, Philadelphia, Pa.
138. Prof. ROBERT BELL (Assistant, Geological Survey of Canada; Secretary, Botanical Society of Canada; Professor of Natural History, Chemistry and Geology, Queen's University), Kingston, Canada West.
139. Prof. W. P. BLAKE (Director, and Professor of Mineralogy, Geology and Mining, College of California), Post Office box 2077, San Francisco, Cal.
140. T. T. BOUVÉ (Curator of Palæontology and Mineralogy, Boston Society of Natural History), Boston, Mass.
141. Prof. P. D. BRADFORD (Professor of Physiology and Pathology, Castleton Medical College), Northfield, Vt.
142. C. G. BREWSTER, No. 16, Tremont street, Boston. *A Dealer.*
143. W. T. BRIGHAM, Boston Society of Natural History, Boston, Mass.
144. C. M. BROWNE, No. 84, John street, New York, N. Y.
145. CHR. C. BROOKS, No. 53, St. Paul street, Baltimore, Md.
146. Prof. GEORGE J. BRUSH (Professor of Mineralogy and Metallurgy, Yale College), New Haven, Ct.
147. STUART M. BUCK, Boston, Mass.
148. A. R. BURTON, Bethlehem, N. H.
149. Prof. JAMES BUSHEE (Curator of Mineralogy, Worcester Society of Natural History), Worcester, Mass.
150. Dr. JOHN CARDEZA, Claymont, Del.
151. SAMUEL R. CARTER, Paris Hill, Oxford Co., Me.

152. Prof. CHARLES F. CHANDLER (Professor of Chemistry, School of Mines, Columbia College), East Forty-ninth street, New York, N. Y.

153. Prof. EDWARD J. CHAPMAN (Professor of Mineralogy and Geology, University College), Toronto, Canada West.

154. ISAAC B. CHOATE, Portland, Me.

155. Rev. A. P. CHUTE, Sharon, Mass.

156. Prof. W. S. CLARK (Professor of Chemistry, Amherst College), Amherst, Mass.

157. JOSEPH A. CLAY, No. 271, South Fifth street, Philadelphia, Pa.

158. Dr. M. H. COATES, Philadelphia, Pa.

159. F. G. COFFIN, Machias, Me.

160. Prof. J. P. COOKE (Professor of Chemistry, Harvard College), Cambridge, Mass.

161. Prof. JAMES D. DANA (Professor of Geology and Mineralogy, Yale College), New Haven, Ct.

162. JOSEPH DELAFIELD, No. 59, Wall street, New York, N. Y.

163. A. DIETY, St. Thomas, West Indies.

164. Prof. ALFRED DU BOIS, Denver, Colorado.

165. E. M. DUNBAR, Springfield, Mass.

166. Rev. E. B. EDDY, Waltham, Mass.

167. Prof. THOMAS EGLESTON (Professor of Mineralogy and Metallurgy, Columbia College), No. 10, Fifth avenue, New York, N. Y.

168. JAMES EIGHTS, Albany, N. Y. *A Dealer.*

169. HENRY ENGELMANN (Mining Engineer), Belleville, Ill.

170. FRANK FAIRBANKS, St. Johnsbury, Vt.

171. CHRISTIAN FEBIYER, Wilmington, Del.

172. M. C. FERNALD, South Levant, Me.

173. Dr. L. FEUCHTWANGER, No. 55, Cedar street, New York, N. Y.

174. GEORGE FISHER, Columbia, Va.

175. Dr. F. A. GENTH, No. 108, Arch street, Philadelphia, Pa.

176. G. K. GILBERT (Assistant, Museum of Prof. Ward), Rochester,

177. Dr. C. A. GOESSMAN, Syracuse, N. Y.          [N. Y.

178. HENRY A. GREEN, Mt. Morris, N. Y.

179. Prof. TRAILL GREEN (Professor of Chemistry, Lafayette College), Easton, Pa.

180. J. J. H. GREGORY, Marblehead, Mass.

181. Prof. GEORGE HADLEY, Buffalo, N. Y.

182. JAMES D. HAGUE, Portage Lake, Mich.

183. Rev. H. F. HARDING, Machias, Me.

184. Prof. LOUIS HARPER (Mining Engineer), No. 1. Rector street, New York, N. Y.

185. JOSHUA P. HASKELL, Marblehead, Mass.

186. THOMAS C. HASKELL, Swampscott, Mass.     [*A Dealer.*
187. CHR. W. A. HERRMANN, No. 607, Broadway, New York, N. Y.
188. CHARLES H. HIGBEE (Curator of Mineralogy, Essex Institute), Salem, Mass.
189. FRANKLIN B. HOUGH, Albany, N. Y.
190. Prof. HENRY HOW (Professor of Chemistry and Natural History, King's College), Windsor, Nova Scotia.
191. WINSLOW J. HOWARD, No. 345, Grand street, New York, N. Y.
192. Prof. F. S. HOYT (Professor of Natural Science, Delaware University), Delaware, Ohio.
193. Prof. O. P. HUBBARD (Professor of Chemistry, Mineralogy and Geology, Dartmouth College), Hanover, N. H.
194. Prof. T. STERRY HUNT (Chemist, Geological Survey of Canada), Montreal, Canada.
195. W. M. HUNTING, Fairfield, Herkimer Co., N. Y.
196. Dr. C. T. JACKSON (Vice President, Boston Society of Natural History), Boston, Mass.
197. W. W. JEFFERIS (Curator, Chester County Cabinet of Natural Science), Westchester, Pa.
198. JOHN JENKINS, Monroe, Orange Co., N. Y.     *A Dealer.*
199. Prof. S. W. JOHNSON (Professor of Agriculture and Analytical Chemistry, Yale College), New Haven, Ct.
200. Prof. JOHN JOHNSTON (Professor of Natural Science, Wesleyan University), Middletown, Ct.
201. Rev. A. B. KENDIG, Marshalltown, Marshall Co., Iowa.
202. JAMES P. KIMBALL, No. 33, Wall street, and No. 40, St. Mark's Place, New York, N. Y.
203. A. C. KLINE, Philadelphia, Pa.     *A Dealer.*
204. W. J. KNOWLTON, Rockport, Mass.
205. EDWARD KOCH, Toledo, Ohio.
206. JOSIAH LADD, Littleton, N. H.
207. Prof. GEORGE LAWSON (Professor of Chemistry Dalhousie College), Halifax, Nova Scotia.
208. ISAAC LEA (Vice President, American Philosophical Society), No. 1622, Locust street, Philadelphia, Pa.
209. ELIAS LEWIS (Chairman of Committee on Natural History, Long Island Historical Society), No. 16, Court street, Brooklyn, N. Y.
210. Prof. A. LITTON (Professor of Chemistry, St. Louis Medical School and Washington University), St. Louis, Mo.
211. JOHN F. LORD, Ellsworth, Me.
212. Prof. O. C. MARSH, Yale College, New Haven, Ct.
213. Prof. JOHN P. MARSHALL (Professor of Chemistry, Mineralogy and Geology, Tufts College), College Hill, Mass.

214. ISAAC C. MARTINDALE (Director, Byberry Philosophical Society), Byberry, Pa.
215. G. F. MATTHEW (Curator, Natural History Society of St. John), St. John, New Brunswick.
216. Hon. ANSON S. MILLER, Rockport, Ill.
217. GIDEON E. MOORE (Curator of Mineralogy, California Academy of Natural Sciences), San. Francisco, Cal.
218. J. V. C. NELLIS, Auburn, N. Y.
219. Prof. J. G. NORWOOD (Professor of Natural Sciences and Natural Philosophy, Missouri State University), Columbia, Boone
220. ALBERT ORDWAY, Post Office box 174, Richmond, Va. [Co., Mo.
221. Rev. JAMES ORTON (Teacher of Natural Sciences, Rochester University), Rochester, N. Y.
222. J. D. PARKER, Steuben, Me.
223. LEROY C. PARTRIDGE, Seneca Falls, N. Y.
224. FRANK PEASLEY, Burlington, Iowa.
225. Prof. MAURICE PERKINS (Professor of Chemistry, Union College), Schenectady, N. Y.
226. Prof. ROBERT PETER (Professor of Chemistry and Experimental Philosophy, Kentucky University), Lexington, Ky.
227. STEPHEN D. POOLE, Lynn, Mass.
228. R. A. RIDEOUT, Garland, Me.
229. W. T. ROEPPER, Bethlehem, Pa.
230. Prof. OREN ROOT (Professor of Geology and Mineralogy, Hamilton College), Clinton, Oneida Co., N. Y.
231. J. G. SANBORN, Cherryfield, Me.
232. E. SEYMOUR, 52, Beekman street, New York, N. Y. *A Dealer.*
233. Dr. WM. SHARSWOOD, Philadelphia, Pa. *Cerium Minerals.*
234. JAMES M. SHAW, South Waterford, Me.
235. Prof. C. U. SHEPARD (Professor of Natural History, Amherst College), Amherst, Mass.
236. Prof. BENJ. SILLIMAN (Professor of Chemistry, Yale College), New Haven, Ct.
237. JOHN P. SIMONS, Philadelphia, Pa.
238. Dr. J. LAWRENCE SMITH, Louisville, Ky.
239. JOHN MILTON SMITH, No. 18, Wall street, New York, N. Y.
240. Prof. CHAS. S. STONE (Professor of Physics and Chemistry, Cooper Institute), New York, N. Y.
241. D. C. STONE, Marysville, Cal.
242. A. P. STUART, Lawrence Scientific School, Cambridge, Mass.
243. WM. H. STURBRIDGE, No. 93, William street, New York, N. Y.
244. Prof. SANBORN TENNEY (Professor of Natural History, Vassar Female College, Poughkeepsie, N. Y.

245. Dr. N. T. TRUE (Editor of Maine Farmer), Bethel, Me.
246. Dr. CHAS. A. TUFTS, Dover, N. H.
247. WM. S. VAUX (Vice President and Curator, Academy of Natural Sciences of Philadelphia), No. 1700, Arch street, Philadelphia, Pa.
248. Prof. A. E. VERRILL (Professor of Zoölogy, Yale College), New Haven, Ct.
249. C. F. WADSWORTH (Curator of Mineralogy, Buffalo Society of Natural Sciences), Buffalo, N. Y.
250. Miss L. E. WALKER (Assistant Curator of Mineralogy, Worcester Society of Natural History), Worcester, Mass.
251. Prof. HENRY A. WARD (Professor of Natural Sciences, Rochester University), Rochester, N. Y.
252. JOHN W. WARD, Salem, Salem Co., N. J.
253. W. E. WELLINGTON, Dubuque, Iowa.
254. CHAS. B. WHITING (Assistant Curator of Mineralogy, Worcester Society of Natural History), Worcester, Mass.
255. S. F. WHITNEY, Brooklyn, N. Y.
256. CHAS. P. WILLIAMS, No. 158, Walnut street, Philadelphia, Pa.
257. Dr. S. C. WILLIAMS, Silver Springs, Lancaster Co., Pa. *Local.*
258. HENRI N. WOODS, Rockport, Mass.

## METALLURGY.

259. OSCAR D. ALLEN, Camden, N. J.
260. WILLIAM ASHBURNER, San Francisco, Cal.
261. Prof. JAMES C. BOOTH, U. S. Mint, Philadelphia, Pa.
262. Prof. GEORGE J. BRUSH (Professor of Mineralogy and Metallurgy, Yale College), New Haven, Ct.
263. Prof. THOMAS EGLESTON (Professor of Mineralogy and Metallurgy, Columbia College), No. 10, Fifth avenue, New York, N. Y.
264. Dr. F. A. GENTH, No. 108, Walnut street, Philadelphia, Pa.
265. Prof. N. P. HILL, Providence, R. I.
266. JAMES T. HODGE, Newburg, N. Y.
267. HENRY JANIN, San Francisco, Cal.
268. LOUIS JANIN, Virginia City, Nevada.
269. EDWARD N. KENT, U. S. Assay Office, New York, N. Y.
270. GUIDO KUESTEL, San Francisco, Cal.
271. GIDEON E. MOORE, Virginia City, Nevada.
272. Prof. JOHN TORREY, (U. S. Assayer), New York, N. Y.
273. Prof. J. D. WHITNEY (State Geologist of California), San Francisco. Cal., and Northampton, Mass.
274. HENRY WURTZ, New York, N. Y.

## PALÆONTOLOGY.

275. Prof. Louis Agassiz (Professor of Zoölogy and Geology, Harvard University; Director and Curator Museum of Comp. Zoölogy), Cambridge, Mass. *General.*

276. A. E. R. Agassiz (Assistant, Museum of Comp. Zoölogy), Cambridge, Mass. *Radiates and Articulates.* Special, *Echinoderms.*

277. Dr. O. P. Baer, Richmond, Ind. *Local.*

278. Henry M. Bannister, Evanston, Ill. *North American.*

279. James E. Barden, Gravesville, Herkimer Co., N. Y. *Local.*

280. Rev. Joseph S. Barris, Sheenwater, Grand Island, N. Y. *Local.*

281. Prof. Wm. H. Barris, Davenport, Iowa. *Crinoids.*

282. J. S. Batterson, Hartford, Ct. *A Dealer.*

283. Prof. Robert Bell (Assistant, Geological Survey of Canada; Secretary, Botanical Society of Canada; Professor of Natural History, Chemistry and Geology, Queen's University), Kingston, Canada West. *North American.*

284. E. Billings (Palæontologist, Geological Survey of Canada), Montreal, Canada. *General.*

285. T. T. Bouvé (Curator of Palæontology and Mineralogy, Boston Society of Natural History), Boston, Mass. *General Collection.*

286. Frank H. Bradley, New Haven, Ct. *North American.*

287. C. G. Brewster, No. 16, Tremont st., Boston, Mass. *A Dealer.*

288. George C. Brown (Curator and Treasurer, Burlington County Lyceum of History and Natural History), Mount Holly, N. J. *Green Sand fossils of New Jersey.*

289. S. T. Carley, Cincinnati, Ohio. *Local.*

290. L. B. Case, Richmond, Ind. *North American.*

291. Charles A. Chase (Librarian and Assistant Curator of Palæontology, Worcester Society of Natural History), Worcester, Mass. *General Collection.*

292. T. Apoleon Cheney (Librarian, Georgic Library), Havana, N. Y. *Local.*

293. David Christy, Cincinnati, Ohio. *Local.*

294. C. Cobb, Buffalo, N. Y. *Local.*

295. T. A. Conrad, Academy of Natural Sciences, Philadelphia, Pa. *North American.*

296. Rev. Sylvester Cowles, West Randolph, Cattaraugus Co., N. Y. *Local.*

297. Prof. James D. Dana (Professor of Mineralogy and Geology, Yale College), New Haven, Ct. *General.*

298. Prof. J. W. Dawson (Principal, McGill University), Montreal, Canada. *British North America.*

299. THOMAS DEVINE (Surveyor in Chief, Upper Canada), Ottawa City, Canada. *North American.*
300. ANDREW DICKSON, Kingston, Canada West. *North American.*
301. CHAS. B. DYER, Cincinnati, Ohio. *Local.*
302. Dr. N. N. ELROD, Little Orleans, Ind. *Local.*
303. W. M. GABB (Palæontologist, California State Geological Survey; Curator of Palæontology, California Academy of Natural Sciences), San Francisco, Cal. *North American.*
304. JOHN GEBHARD, Jr., Schoharie, N. Y. *Local.*
305. G. R. GILBERT (Assistant, Museum of Prof. Ward), Rochester, N. Y. *General Collection.*
306. HENRY A. GREEN, Mount Morris, N. Y. *Local.*
307. B. J. HALL, Burlington, Iowa. *Crinoids. A Collector.*
308. Prof. JAMES HALL (State Geologist of New York, Iowa and Wisconsin), Albany, N. Y. *General.*
309. RICHARD HAMANT (Curator of Palæontology, Worcester Society of Natural History), Worcester, Mass. *General Collection.*
310. C. FRED. HARTT, Museum of Comp. Zoölogy, Cambridge, Mass. *North American.*
311. Prof. F. V. HAYDEN (Professor of Geology and Mineralogy, University of Pennsylvania), Smithsonian Institution, Washington, D. C. *North American.*
312. E. W. HILGARD (State Geologist of Mississippi; Professor of Chemistry, University of Mississippi), Oxford, Miss. *North American.*
313. Rev. FREDERIC HUIDEKOPER, Meadville, Pa. *Local.*
314. GEORGE W. HOLDEN, Dayton, Ohio. *Local.*
315. Prof. FRANCIS S. HOLMES, College of Charleston, Charleston, S. C. *South Carolinian.*
316. THEODORE HOWLAND (Secretary and Curator of Palæontology, Buffalo Society of Natural Sciences), Buffalo, N. Y. *General Collection.*
317. Dr. E. W. HUBBARD, Tottenville, Staten Island, N. Y. *Local.*
318. W. M. HUNTING, Fairfield, Herkimer Co., N. Y. *Local.*
319. ALPHEUS HYATT (Curator of Mollusca, Boston Society of Natural History), Boston, Mass. *Cephalopods.*
320. U. P. JAMES, Cincinnati, Ohio. *United States.*
321. JOHN JENKINS, Monroe, Orange Co., N. Y. *A Dealer.*
322. Col. EZEKIEL JEWETT, Utica, N. Y. *New York State.*
323. W. D. JOHNSON, Cochocton, Ohio. *Local.*
324. Dr. JAMES KNAPP, Louisville, Ky. *Local.*
325. Dr. ALBERT C. KOCH, St. Louis, Mo. *North American.*
326. ISAAC KENLEY, Richmond, Ind. *Local.*

327. ISAAC LEA (Vice President, American Philosophical Society), No. 1622, Locust street, Philadelphia, Pa. *North American.*
328. Prof. JOSEPH LEIDY (Professor of Anatomy, University of Pennsylvania; Curator, Academy of Natural Sciences of Philadelphia), No. 1302, Filbert street, Philadelphia, Pa. *General.*
329. Dr. A. M. LEONARD, Lockport, N. Y. *Local.*
330. Prof. LEO LESQUEREUX, Columbus, Ohio. *Plants.*
331. Rev. SAMUEL LOCKWOOD, Keyport, N. J. *Cretaceous fossils of New Jersey and Devonian Plants of New York.*
332. S. L. LYON, Kanawha Court House, Va. *Local.*
333. Dr. R. P. MANN, Milford, Ohio. *Devonian Fishes.*
334. Prof. O. C. MARSH (Professor of Palæontology, Sheffield Scientific School, Yale College,) New Haven, Ct. *General.*
335. CHARLES D. MARSHALL (Corresponding Secretary, Buffalo Society Natural Science), Buffalo, N. Y. *Local.*
336. JOHN E. MARSHALL, Buffalo, N. Y. *Local.*
337. E. MATHEWSON, Martinez, Cal. *Local.*
338. Prof. J. H. McCHESNEY, Jacksonville, Ill. (U. S. Consul at New Castle on-Tyne). *North American.*
339. F. B. MEEK, Smithsonian Institution, Washington, D. C. *General.*
340. Prof. W. D. MOORE, Irwin's Station, Pa. *Carboniferous.*
341. Prof. J. S. NEWBERRY (Professor of Geology, Columbia College), New York, N. Y. *North American.* Special, *Fishes and Plants.*
342. W. H. NILES, New Haven, Ct. *North American.* Special, *Crinoids.*
343. SIDNEY A. NORTON, Cleveland, Ohio. *Local.*
344. Prof. J. G. NORWOOD (Professor of Natural Sciences and Natural Philosophy, Missouri State University), Columbia, Boon Co., Mo. *North American.*
345. J. KELLY O'NEALE, Lebanon, Ohio. *Local.*
346. ALBERT ORDWAY, Post Office box 174, Richmond, Va. *Local.*
347. Rev. JAMES ORTON (Teacher of Natural Sciences, Rochester University), Rochester, N. Y. *Local.*
348. Dr. A. S. PACKARD, jr. (Curator of Crustacea, Boston Society of Natural History; Curator of Radiata and Articulata, Essex Institute), Salem, Mass. *New England Drift.*
349. FRANK PEASELY, Burlington, Iowa. *Local.*
350. ALFRED POOLE, Halifax, Nova Scotia. *Local.*
351. Prof. J. W. POWELL, Bloomington, Ill. *North American.*
352. Hon. H. S. RANDALL, Cortland Village, N. Y. *Local.*

353. Dr. EDMUND RAVENEL, Charleston, S. C. *Invertebrates.*
354. AUGUSTE RÉMOND, San Francisco, Cal. *Local.*
355. A. B. RICHMOND, Meadville, Pa. *North American.*
356. Dr. CARL ROMINGER (Curator, Museum of the Michigan State University), Ann Arbor, Mich. *North American.*
357. HENRY ROUSSEAU, Troy, N. Y. *Mollusks.*
358. WM. L. RUST, Trenton Falls, N. Y. *Trilobites.*
359. S. H. SCUDDER (Secretary, Librarian and Curator of Entomology, Boston Society of Natural History), Boston, Mass. *Insects.*
360. D. H. SHAFFER, Cincinnati, Ohio. *A Collector.*
361. N. S. SHALER (Assistant, Museum of Comp. Zoölogy), Cambridge, Mass. *Mollusks.* Special, *Brachiopods.*
362. Dr. B. F. SHUMARD, St. Louis, Mo. *North American.*
363. O. H. ST. JOHN, Waterloo, Iowa. *Local.*
364. Prof. G. C. SWALLOW (State Geologist of Kansas and Missouri), Columbia, Boone Co., Mo. *North American.*
365. GEORGE W. TAYLOR, Pulaski, N. Y. *Local.*
366. Dr. O. THIEME, Burlington, Iowa. *Crinoids. A Collector.*
367. Dr. JOHN B. TRASK, San Francisco, Cal. *Local.*
368. Prof. A. E. VERRILL (Professor of Zoölogy, Yale College), New Haven, Ct. *Corals.*
369. CHARLES WACHSMUTH, Burlington, Iowa. *Crinoids.*
370. Prof. HENRY A. WARD (Professor of Natural Sciences, Rochester University), Rochester, N. Y. *General Collection.*
371. CHARLES M. WHEATLEY, Phœnixville, Pa., and No. 42, Pine street, New York, N. Y. *Mesozoic.*
372. Dr. CHARLES A. WHITE, Iowa City, Iowa. *North American.*
373. R. P. WHITFIELD, Albany, N. Y. *North American.*
374. Prof. ALEXANDER WINCHELL (Professor of Natural History, Michigan State University), Ann Arbor, Mich. *North American.*

## PHYSICAL GEOGRAPHY.

375.  Bv't Brig. Gen. HENRY L. ABBOTT, U. S. Engineers. *North American.*
376.  Prof. A. D. BACHE (Superintendent, United States Coast Survey), Washington, D. C. *North American.*
377.  G. H. BAGWELL (Assistant, United States Coast Survey), Washington, D. C. *North American.*
378.  Brig. Gen. JOHN G. BARNARD, U. S. Engineers. *North American.*
379.  Col. E. G. BECKWITH, U. S. Army. *North American.*
380.  Prof. WM. P. BLAKE (Professor of Mineralogy, Geology and Mining, College of California; Geologist of the California State Board of Agriculture), Post office box 2077, San Francisco, Cal. *Californian.*
381.  Prof. WM. H. BREWER (Professor of Agriculture, Yale College), New Haven, Ct. *California Mountains.*
382.  Brig. Gen. JAMES H. CARLETON, U. S. Army. *North American.*
383.  Prof. EDWARD J. CHAPMAN (Professor of Mineralogy and Geology, University College), Toronto, Canada West. *North American.*
384.  Maj. Gen. S. W. CRAWFORD, U. S. Army. *North American.*
385.  Lieut. Col. OSBORNE CROSS, U. S. Army. *North American.*
386.  Brig. Gen. RICHARD DELAFIELD, Chief Engineer, U. S. Army. *North American.*
387.  F. W. DORR (Assistant, United States Coast Survey), Washington, D. C. *North American.*
388.  Maj. Gen. W. H. EMORY, U. S. Army. *North American.*
389.  Dr. GEORGE ENGELMANN, St. Louis, Mo. *North American.*
390.  N. S. FINNEY (Assistant, United States Coast Survey), Washington, D. C. *North American.*
391.  JOHN C. FREMONT, St. Louis, Mo. *North American.*
392.  Prof. DANIEL C. GILMAN (Professor of Physical Geography, Yale College), New Haven, Ct. *General.*
393.  Prof. ARNOLD GUYOT (Professor of Geology and Physical Geography, College of New Jersey), Princeton, N. J. *General.*
394.  Prof. JOSEPH HENRY (Secretary, Smithsonian Institution), Washington, D. C. *General.*
395.  J. E. HILGARD (Assistant, United States Coast Survey), Washington, D. C. *North American.*

NOTE. Army Officers are best reached by addressing them care of the ADJUTANT GENERAL, U. S. A., Washington, D. C.

396. Prof. HENRY Y. HIND (Professor of Physical and Natural Sciences, Trinity College), Toronto, Canada West. *North American.*

397. JOHN GEORGE HODGINS (Deputy Superintendent of Education for Upper Canada), Education Office, Toronto, Canada West. *General.*

398. Maj. Gen. A. A. HUMPHREYS, U. S. Engineers. *North American.*

399. CLARENCE KING (Assistant, California Geological Survey), San Franciscó, Cal. and Irvington, N. Y. *Californian.*

400. J. S. LAWSON (Assistant, United States Coast Survey), Washington, D. C. *North American.*

401. Dr. THOMAS M. LOGAN, Sacramento, Cal. *Californian.*

402. Lieut. Col. JOHN N. MACOMB, U. S. Engineers. *North American.*

403. Col. RANDOLPH B. MARCY, U. S. Army. *North American.*

404. Bv't Lieut. Col. H. E. MAYNADIER, U. S. Army, No. 388, 19th street, Washington, D. C. *North American.*

405. Maj. Gen. GEORGE G. MEADE, U. S. Engineers, Philadelphia, Pa. *North American.*

406. Lieut. Col. N. MICHLER, U. S. Engineers. *North American.*

407. Maj. Gen. JOHN G. PARKE, U. S. Engineers. *North American.*

408. Dr. C. C. PARRY, Davenport, Iowa. *Rocky Mountains.*

409. Maj. Gen. JOHN POPE, U. S. Army. *North American.*

410. Col. W. F. RAYNOLDS, U. S. Engineers. *North American.*

411. C. A. SCHOTT (Assistant, United States Coast Survey), Washington, D. C. *North American.*

412. Lieut. Col. J. H. SIMPSON, U. S. Army. *North American.*

413. Lieut. Col. LORENZO SITGREAVES, U. S. Engineers. *North American.*

414. ALEXANDER S. TAYLOR, Santa Barbara, Cal. *Californian.*

415. Col. GEORGE THOM, U. S. Engineers. *North American.*

416. Bv't Brig. Gen. STEWART VAN VLIET, U. S. Army. *North American.*

417. Maj. Gen. G. K. WARREN, U. S. Engineers. *North American.*

418. WALTER WELLS, Portland, Me. *General.*

419. Prof. J. D. WHITNEY (State Geologist of California), Northampton, Mass., or San Francisco, Cal. *North American.*

420. Major R. S. WILLIAMSON, U. S. Engineers, San Francisco, Cal. *North American.*

421. Lieut. Col. I. C. WOODRUFF, U. S. Engineers. *North American.*

422. Maj. Gen. H. G. WRIGHT, U. S. Engineers. *North American.*

## COMPARATIVE ANATOMY AND PHYSIOLOGY.

423. Prof. HARRISON ALLEN (Professor of Comparative Anatomy, University of Pennsylvania), Philadelphia, Pa.

424. Prof. H. JAMES CLARK (Professor of Natural History, Pennsylvania Agricultural College), Centre Co., Pa, *Invertebrates.*

425. Prof. JOHN C. DALTON (Professor of Physiology and Microscopic Anatomy, Columbia College; Professor of Physiology, College of Physicians and Surgeons), New York, N. Y.

426. Dr. JOHN DEAN, Boston, Mass.

427. Prof. JOHN C. DRAPER (Professor of Analytical Chemistry, University of New York; Professor of Natural History and Physiology, New York Free Academy), New York, N. Y.

428. Prof. AUSTIN FLINT, jr. (Professor of Physiology, Bellevue Hospital), New York, N. Y.

429. Dr. WM. A. HAMMOND, No. 162, West Thirty-fourth street, New York, N. Y.

430. Prof. EDWARD HITCHCOCK (Professor of Hygeine and Physical Education, Amherst College), Amherst, Mass.

431. Dr. JULIUS HOMBERGER (Editor, American Journal of Opthalmology), No. 39, West Twenty-third street, New York, N. Y.

432. Prof. CHRISTOPHER JOHNSON (Professor of Anatomy and Physiology, University of Maryland), Baltimore, Md.

433. Prof. JOSEPH LEIDY (Professor of Anatomy, University of Pennsylvania; Curator, Academy of Natural Sciences of Philadelphia), No. 1302, Filbert street, Philadelphia, Pa.

434. Dr. J. S. LOMBARD (Assistant Professor of Physiology, Medical School of Harvard University), Boston, Mass.

435. Prof. MANLY MILES (Professor of Animal Physiology and Practical Agriculture, State Agricultural College), Lansing, Mich.

436. Dr. S. WEIR MITCHELL, Academy of Natural Sciences, or No. 1332, Walnut street, Philadelphia, Pa.

437. GEORGE SCEVA, Townsend, Mass.      [New York, N. Y.

438. Prof. J. CRESSON STILES (Professor of Physiology, ———————),

439. Dr. HENRY WHEATLAND (Secretary, Treasurer and Curator of Camparative Anatomy, Essex Institute), Salem, Mass.

440. Dr. J. C. WHITE (Curator of Comparative Anatomy and Mammalogy, Boston Society of Natural History), Boston, Mass.

441. Dr. B. G. WILDER(Curator of Herpetology, Boston Society of Natural History ), No. 54, Bowdoin street, Boston, Mass.

442. Dr. RUFUS WOODWARD (President, Worcester Society of Natural History), Worcester, Mass.

443. Prof. JEFFRIES WYMAN (Professor of Anatomy and Physiology, Harvard University; President, Boston Society of Natural History), Cambridge, Mass.

## VEGETABLE PHYSIOLOGY.

444. Prof. S. W. JOHNSON (Professor of Agricultural Chemistry, Yale College), New Haven, Ct.
445. JOHN H. KLIPPART, Columbus, Ohio.
446. S. B. MCMILLAN, East Fairfield, Columbiana Co., Ohio.

## HISTOLOGY.

447. Prof. H. JAMES CLARK (Professor of Natural History, Pennsylvania Agricultural College), Centre Co., Pa.
448. Prof. JOSEPH LEIDY (Professor of Anatomy, University of Pennsylvania; Curator, Academy of Natural Sciences of Philadelphia), No. 1302, Filbert street, Philadelphia, Pa.
449. Prof. J. H. SALISBURY (Professor of Physiology, Histology and Cell Pathology, Charity Hospital Medical College), Cleveland, Ohio.
450. THEODORE A. TELLKAMPF, No. 142, West Fourth street, New York, N. Y.
451. Prof. JEFFRIES WYMAN (Professor of Anatomy and Physiology, Harvard University; President, Boston Society of Natural History), Cambridge, Mass.

## EMBRYOLOGY.

452. A. E. R. AGASSIZ (Assistant, Museum of Comp. Zoölogy), Cambridge, Mass.
453. Prof. LOUIS AGASSIZ (Professor of Zoölogy and Geology, Harvard University; Director and Curator, Museum of Comparative Zoölogy), Cambridge, Mass.
454. Prof. H. JAMES CLARK (Professor of Natural History, Pennsylvania Agricultural College), Centre Co., Pa.
455. Prof. A. E. VERRILL (Professor of Zoölogy, Yale College), New Haven, Ct.
456. Prof. JEFFRIES WYMAN (Professor of Anatomy and Physiology, Harvard University; President, Boston Society of Natural History),Cambridge, Mass.

## MICROSCOPY.

457. Dr. GEORGE S. ALLAN, Newburgh, N. Y.
458. J. W. S. ARNOLD (Librarian, Microscopical Society), New York, N. Y.
459. Prof. L. W. BAILEY (Professor of Chemistry and Natural History, University of New Brunswick), Fredericton, N. B.
460. MOSES Y. BEACH, Louisville, Ky.
461. EDWIN BICKNEL*, Essex Institute, Salem, Mass.
462. Dr. STEPHEN W. BOWLES, Brattleborough, Vt.
463. Prof. W. H. BREWER (Professor of Agriculture, Yale College), New Haven, Ct.
464. Dr. RUFUS K. BROWNE (Corresponding Secretary, American Microscopical Society), New York, N. Y.
465. Prof. H. JAMES CLARK (Professor of Natural History, Pennsylvania Agricultural College), Centre Co., Pa.
466. Prof. ARTHUR S. COPEMAN (Professor of ———, New York College of Veterinary Surgeons), New York, N. Y.
467. CALEB COOKE,* Essex Institute, Salem, Mass.
468. ROBERT DINWIDDIE (Vice President, American Microscopical Society), No. 79, West Twentieth street, New York, N. Y.
469. GEORGE DOBBIN, St. Paul street, Baltimore, Md.
470. T. D'ORÉMIEULX (Treasurer, American Microscopical Society), No. 261, Greene street, New York, N. Y.
471. A. M. EDWARDS (President, American Microscopical Society), No. 49, Jane street, New York, N. Y.
472. JAMES H. EMERTON,* Essex Institute, Salem, Mass.
473. CHRISTIAN FEBIGER, Wilmington, Del.
474. JOSEPH FENTON, Columbus, Ohio.
475. DAVID W. FERGUSON, New York, N. Y.
476. WILLIAM H. GARLICK, Cleveland, Ohio.
477. JNO. E. GAVIT (Vice President, American Microscopical Society), No. 83, West Forty-third street and No. 142, Broadway, New York, N. Y.
478. RICHARD C. GREENLEAF, Boston, Mass.
479. T. F. HARRISON, New York, N. Y.
480. D. S. HINES, Brooklyn, N. Y.
481. Dr. J. H. HINTON (Recording Secretary, American Microscopical Society), New York, N. Y.
482. WILLIAM E. HULBERT, Middletown, Ct.
483. WILLIAM W. HUSK, Brooklyn, N. Y.

484. ALPHEUS HYATT* (Curator of Mollusca, Boston Society of Natural History; Curator of Palæontology and Protozoa, Essex Institute), Salem, Mass.
485. SAMUEL JACKSON (Curator, American Microscopical Society), New York, N. Y.
486. Dr. B. JOY JEFFRIES (Curator of Microscopy, Boston Society of Natural History), Boston, Mass.
487. Prof. CHRISTOPHER JOHNSTON (Professor of Anatomy, Maryland College), Baltimore, Md.
488. Dr. SAMUEL A. JONES, Englewood, N. J.
489. H. F. KING,* Essex Institute, Salem, Mass.
490. Dr. JOHN KING, Ciucinnati, Ohio.
491. Dr. F. W. LEWIS, Philadelphia, Pa.
492. THOMAS H. MCALLISTER, No. 49, Nassau street, New York, N. Y.
493. Prof. OGDEN N. ROOD (Professor of ————, Columbia College), New York, N. Y.
494. Dr. W. V. V. ROSA, Watertown, N. Y.
495. LEWIS M. RUTHERFORD, No. 175, Second Avenue, New York, N. Y.
496. H. F. SHEPARD,* Essex Institute. Salem, Mass.
497. Prof. HAMILTON L. SMITH (Professor of Natural History and Astronomy, Kenyon College), Gambier, Ohio.
498. T. W. STARR, Philadelphia, Pa.
499. CHARLES STODDER, No. 75, Kilby street, Boston, Mass.
500. CORNELIUS VAN BRUNT, Fishkill-on-the-Hudson, N. Y.
501. BENJAMIN WEBB, jr.* Essex Institute, Salem, Mass.
502. Dr. M. C. WHITE, New Haven, Ct.
503. Dr. T. G. WORMLEY, Columbus, Ohio.

---

*These gentlemen are Curators of the Microscopical Section of the Essex Institute.

# BOTANY.

504. Prof. WM. E. A. AIKIN (Professor of Chemistry, Baltimore University), Baltimore, Md. *North American.*

505. W. P. ALCOTT, Andover, Mass. *Local.*

506. Miss. LIZZIE B. ALLEN, Davenport, Iowa. *Local.*

507. Dr. T. F. ALLEN, New York, N. Y. *North American.*

508. Dr. C. L. ANDERSON, Santa Cruz, Cal. *Nevada.*

509. Dr. T. L. ANDREWS, Niagara Falls, N. Y. *Californian.*

510. Dr. T. L. ANTISELL, U. S. Army. *Pacific Rail Road Route.*

511. C. F. AUSTIN, Closter, Bergen Co., N. J. *North American.* Special, *Hepaticæ.*

512. E. P. AUSTIN, Nautical Almanac Office, Washington, D. C. *Michigan.*

513. AUSTIN BACON, Natick, Mass. *Local.*

514. Dr. M. M. BAGG, Utica, N. Y. *Local.*

515. Prof. L. W. BAILEY (Professor of Chemistry and Natural History, University of New Brunswick), Fredericton, N. B. *Local.*

516. VINCENT BARNARD, Kennett Square, Chester Co., Pa. *Local.*

517. Dr. JACOB BARRATT, Middletown, Mass. *Local.* Special, *Salices.*

518. Dr. JOSEPH BATES, New Lebanon Spa, Columbia Co., N. Y. *Local.*

519. WM. J. BEAL (Teacher of Natural Sciences, Young Ladies' Collegiate Institute), Union Springs, Cayuga Co., N. Y. *North American.*

520. Miss M. E. BEAUCHAMP, Skaneateles, N. Y. *Local.*

521. Rev. M. W. BEAUCHAMP, King's Ferry, Cayuga Co., N. Y. *Local.*

522. M. S BEBB, Rockford, Ill. *North American.*

523. JAMES L. BENNETT, Providence, R. I. *Local.*

524. Dr. JACOB BIGELOW, Boston, Mass. *North American.*

525. Dr. J. M. BIGELOW, Detroit, Mich. *North American.*

526. B. BILLINGS, Ottawa City, Canada West. *Canadian.*

527. Rev. JOSEPH BLAKE, Gilmantown, N. H. *Local.*

528. H. G. BLOOMER, San Francisco, Cal. *California and Nevada.*

529. Prof. H. N. BOLANDER (Botanist, California State Survey; Curator of Botany, California Academy of Natural Science), San Francisco, Cal. *Californian.* Special, *Cryptogames and Glumaceæ.*

530. Dr. C. M. BOOTH, Rochester, N. Y. *Local.*

531. WM. BOOTT, Boston, Mass. *North American.* Special, *Ferns, Grasses and Carices.*

532. Dr. E. D. Bostwick, Litchfield, Ct. *Local.*
533. Mrs. —— Bowen, Skaneateles, N. Y. *Local.*
534. William Bower, No. 53 Fulton street, New York, N. Y. *North American Ferns and Orchids under cultivation.*
535. Dr. Frederick Brendel, Peoria, Ill. *North American.*
586. Prof. Wm. H. Brewer (Professor of Agriculture, Yale College), New Haven, Ct. *North American.* Special, *Californian.*
537. Dr. Robert Bridges (President, Academy of Natural Sciences of Philadelphia), Philadelphia, Pa. *Local.*
538. Garland C. Broadhead, Pleasant Hill, Mo. *Local.*
539. The Abbé Ovide Brunet (Professor of Botany, Université Laval; President, Entomological Society of Canada, Quebec Branch), Quebec, Canada. *North American.* Special, *Canadian.*
540. —— Buchanan, New York, N. Y. *Cultivated Plants and Local.*
541. Robert Buchanan, Cincinnati, Ohio. *Local.*
542. S. B. Buckley, ——, Texas. *Local.*
543. Dr. F. J. Bumstead, No. 162 West Twenty-third street, New York, N. Y. *North American.*
544. Isaac Burk, Philadelphia, Pa. *Local.*
545. Wm. M. Canby, No. 834 Market street, Wilmington, Del. *North American.*
546. Dr. C. A. Canfield, Monterey, Cal. *Californian.*
547. Dr. Joseph Carson, Philadelphia, Pa. *Medical Botany and Local.*
548. Dr. Francis Carter, Columbus, Ohio, *Local.*
549. Prof. J. Lang Cassells (Professor of Chemistry and Toxicology, Western Reserve College), Hudson, Ohio. *North American.*
550. Prof. P. A. Chadbourne (President, Massachusetts State Agricultural College), Amherst, Mass. *North American.*
551. Dr. A. W. Chapman, Apalachicola, Fla. *North American.*
552. Rev. J. W. Chickering, Exter, N. H. *North American.*
553. Dr. —— Clapp, New Albany, Ind. *Western States.*
554. Dr. Daniel Clark (President, Flint Scientific Institute), Flint, Mich. *Local.*
555. Prof. H. James Clark (Professor of Natural History, Pennsylvania Agricultural College), Centre Co., Pa. *North American.*
556. James H. Clark, Newport, R. I. *Carboniferous Plants.*
557. Miss Mary H. Clark, Ann Arbor, Mich. *Local.*
558. Prof. W. S. Clark (Professor of Chemistry, Amherst College), Amherst, Mass. *North American.*
559. Hon. George W. Clinton (President, Buffalo Society of Natural Science), Buffalo, N. Y. *North American.*

560. A. COMMONS, Centreville, Del. *Local.*
561. Dr. —— COOLEY, Utica, Mich. *Local.*
562. Dr. J. G. COOPER (Zoölogist, California State Survey and California State Board of Agriculture; Curator of Zoölogy, California Academy of Natural Science), Santa Cruz, Cal. *Pacific Coast.*
563. Dr. I. COURT, Port of Spain, Trinidad, W. I. *West Indian.*
564. Dr. E. S. CROSIER, New Albany, Ind. *Local.*
565. Rev. Dr. M. A. CURTIS, Hillsborough, N. C. *North American.* Special, *Fungi.*
566. A. H. CURTISS, Liberty, Bedford Co., Va. *Local.*
567. CHRISTIAN DAHL, St. Croix, W. I. *West Indian.*
568. Dr. JOHN DARBY, Auburn, Ala. *North American.*
569. Dr. J. DARRACH, Germantown, Pa. *Local.*
570. DAVID F. DAY, Buffalo, N. Y. *Local.*
571. Dr. D. V. DEAN, St. Louis, Mo. *Local.*
572. W. W. DENSLOW, Inwood Station, Hudson River R. R.; Post-office address, Station "N," New York, N. Y. *North American.*
573. Prof. CHESTER DEWEY (Professor of Chemistry and Natural History, University of Rochester), Rochester, N. Y. *North American.* Special, *Carices.*
574. Dr. ELIAS DIEFFENBAUGH, Clinton street, Philadelphia, Pa. *North American.*
575. Mrs. WM. E. DOGGETT, Chicago, Ill. *Local.*
576. Dr. EDWARD DORSCH, Monroe, Mich. *Local.*
577. A. T. DRUMMOND, Montreal, Canada. *Canadian.*
578. ELIAS DURAND, Philadelphia, Pa. *North American.*
579. Miss A. B. EARLE (Assistant Curator of Botany, Worcester Society of Natural History), Worcester, Mass. *Local.*
580. Prof. D. C. EATON (Professor of Botany, Yale College), New Haven, Ct. *General.* Special, *Filices.*
581. ARTHUR M. EDWARDS, No. 49 Jane street, New York, N. Y. *Diatomaceæ.*
582. SAMUEL E. ELMORE, Hartford, Ct. *Local.*
583. GEORGE B. EMERSON, Boston, Mass. *New England.* Special, *Forest Trees.*
584. RUSH. EMERY, Tipton, Iowa. *Local.*
585. FRANCIS E. ENGELHARDT (Professor of Chemistry, St. Francis College), No. 49 West Fifteenth st., New York, N. Y. *Local.*
586. Dr. GEORGE ENGELMANN (President, St. Louis Academy of Science), St. Louis, Mo. *General.*
587. Prof. JACOB ENNIS, Philadelphia, Pa. *Local.*

588. Miss —— Errington, New York, N. Y. *Californian.*
589. Dr. O. Everett, Dixon, Lee Co., Ill. *Local.*
590. Augustus Fendler, Allenton, St. Louis Co., Mo. *New Mexico and South American.*
591. I. Foote, Detroit, Mich. *Local.*
592. Dr. E. Foreman, Catonsville, Md. *Local.*
593. Rev. James Fowler, Richibucto, N. B. *Local.*
594. Samuel P. Fowler (Vice President, Essex Institute), Danvers, Mass. *Forest Trees of New England.*
595. J. Q. A. Fritchey, St. Louis, Mo. *Local.*
596. Charles C. Frost, Brattleborough, Vt. *Local.* Special. *Lichens and Fungi.*
597. Dr. C. C. F. Gay (Curator of Botany, Buffalo Society of Natural Science), Buffalo, N. Y. *Local.*
598. George Gibbs, Washington, D. C. *Oregon and Washington Terr.*
599. Dr. George L. Goodale (Curator of Botany, Portland Society of Natural History), Portland, Me. *North American.*
600. Hon. John S. Gould, Hudson, N. Y. *Local.*
601. Prof. Asa Gray (Professor of Botany, Harvard University: President, American Academy of Arts and Sciences), Cambridge, Mass. *General.* Special, *North American.*
602. Dr. C. Green, Homer, N. Y. *Local.*
603. Prof. Traill Green (Professor of Chemistry, Lafayette College), Easton, Pa. *North American.*
604. Thomas A. Greene, New Bedford, Mass. *North American.*
605. Elihu Hall, Athens, Menard Co., Ill. *North American.*
606. G. P. Harbour, Oskaloosa, Iowa. *Rocky Mountains.*
607. Fielden Hartley, Alton, Ill. *Local.*
608. Clark C. Haskins, New Albany, Ind. *Local.*
609. Prof. F. V. Hayden (Professor of Geology and Mineralogy, University of Pennsylvania), Philadelphia, Pa. *Upper Missouri.*
610. Dr. G. W. Hazletine, Jamestown, N. Y. *Local.*
611. Dr. E. P. Healey, Medina, N. Y. *Local.*
612. E. W. Hervey, New Bedford, Mass. *Local.*
613. Prof. E. W. Hilgard (Professor of Chemistry and Mineralogy, University of Mississippi), Oxford, Miss. *North American.*
614. Dr. Theo. C. Hilgard, St. Louis, Mo. *Local.* Special, *Fungi and Algæ.*
615. Prof. William Hincks (Professor of Natural History, University College; Editor, Canadian Journal of Industry, Science, and Art), Toronto, C. W. *General.* Special, *Canadian.*
616. Hon. G. H. Hollister, Litchfield, Ct. *Local.*

617. E. S. HOLMES, Wilson, Niagara Co., N. Y. *Local.*
618. I. F. HOLTON, Medford, Mass. *New Grenada and Local.*
619. JOSHUA HOOPES, Westchester, Pa. *Local.*
620. Dr. ASA HORR, Dubuque, Iowa. *Local.*
621. FRANKLIN B. HOUGH, Albany, N. Y. *North American.*
622. Prof. HENRY HOW (Professor of Chemistry and Natural History, King's College), Windsor, N. S. *Local.*
623. WINSLOW J. HOWARD, No. 345 Grand street, New York, N. Y. *Rocky Mountains.*
624. Dr. ELLIOTT C. HOWE, Troy, N. Y. *Local.*
625. Prof. JAMES HUBBERT (Professor of Natural Sciences, St. Francis College), Richmond, C. E. *Canadian.* Special, *Fungi.*
626. Dr. A. T. HUDSON, Lyons, Clinton Co., Iowa. *Local.*
627. Dr. G. W. HULSA, Natchez, Miss. *California and Florida.*
628. GEORGE HUNT, Providence, R. I. *Local.*
629. ROBERT INGRAHAM, New Bedford, Mass. *North American Cryptogamia.*
630. HALLIDAY JACKSON, Westchester, Chester Co., Pa. *Local.*
631. Prof. THOMAS P. JAMES (Professor of Botany, Pennsylvania Horticultural Society), No. 400 South Ninth street, Philadelphia, Pa. *North American.* Special, *Musci.*
632. Dr. H. A. JOHNSON, Chicago, Ill. *Local.*
633. Dr. A. KELLOGG (Librarian, Californian Academy of Natural Science), San Francisco, Cal. *Pacific Coast of America.*
634. C. KESSLER, Reading, Pa. *Local.*
635. JOHN KIRKPATRICK (Secretary, Academy of Natural Sciences of Cleveland; Secretary, Cleveland Horticultural Society, Cleveland, Ohio. *Local.*
636. Dr. JARED P. KIRTLAND, East Rockport, Ohio; Post-office address, Cleveland, Ohio. *Local.*
637. Dr. P. D. KNIESKERN, Shark River, N. J. *Local.*
638. HENRY KREBS, St. Thomas, W. I. *West Indian.*
639. THURE KUMLIEN, Busseyville Post-office, via Albion, Wis. *Wisconsin.*
640. Hon. I. A. LAPHAM (President, Wisconsin Historical Society), Milwaukee, Wis. *North American.*
641. Miss S. L. LAWRENCE (Curator of Botany, Worcester Society of Natural History), Worcester, Mass. *Local.*
642. Prof. GEORGE LAWSON (Professor of Chemistry, Dalhousie College), Halifax, N. S. *North American.*
643. W. H. LEGGETT, No. 224 Tenth street, New York, N. Y. *Local.*

644. Prof. LEO LESQUEREUX, Columbus, Ohio. *General.* Special, *Musci and Fossil Plants.*
645. FERDINAND LINDHEIMER, New Braunfels, Texas. *Texan.*
646. Dr. GEORGE LITTLE (Mississippi State Geologist), Oxford, Miss. *South Western States.*
647. Rev. SAMUEL LOCKWOOD, Keyport, N. J. *Devonian Plants of New York.*
648. Rev. J. E. LONG, Hublersburg, Centre Co., Pa. *Local.*
649. H. B. LORD, Ithica, N. Y. *Local.*
650. Dr. STARLING LOVING, Columbus, Ohio. *Local.*
651. J. R. LOWRIE, Olive Post-office, Pa. *Local.*
652. JOHN MACOUN, Belleville, C. W. *Canadian.*
653. HORACE MANN (Curator of Botany, Boston Society of Natural History), Cambridge, Mass. *General.*
654. W. T. MARCH, Spanishtown, Jamaica. *West Indian.*
655. ISAAC C. MARTINDALE (Director, Byberry Philosophical Society), Byberry, Pa. *North American.*
656. Dr. JOSEPH C. MARTINDALE, No. 918 North Twelfth street, Philadelphia, Pa. *Local.*
657. R. MATTHEW, No. 93 Princess street, St. John, N. B. *Local.*
658. Dr. S. B. MEAD, Augusta, Hancock Co., Ill. *Local.*
659. THOMAS MEEHAN (Corresponding Secretary, Pennsylvania Horticultural Society; Editor, Gardener's Monthly), Germantown, Pa. *General.* Special, *Horticultural.*
660. Dr. EZRA MICHENER, Avondale, Chester Co., Pa. *North American.* Special, *Fungi.*
661. Hon. ANSON S. MILLER, Rockford, Ill. *Local.*
662. CH. MOHR, Mobile, Alabama. *Local.*
663. Prof. JOSEPH MOORE (Professor of Natural History, Earlham College), Richmond, Ind. *Local.*
664. Prof. W. D. MOORE, Irwin's Station, Pa. *Local.*
665. Dr. SEBASTIAN ALFREDO DE MORALES, Calle de Velarde, No. 5, Matanzas, Cuba. *Cuban.*
666. Miss E. S. MORSE (Assistant Curator of Botany, Worcester Society of Natural History), Worcester, Mass. *Local.*
667. Miss M. E. B. MORTON, Rockford, Winnebago Co., Ill. *Local.*
668. WILLIAM MUIR, Fox Creek Post-office, St. Louis Co., Mo. *Horticultural.*
669. Prof. J. S. NEWBERRY (Professor of Geology, Columbia College), New York, N. Y. *North American.* Special, *Fossil Plants.*
670. S. T. OLNEY, Providence, R. I. *Local.*

671. Dr. J. G. ORTON, Binghamton, N. Y. *Local.*
672. Dr. HORACE M. PAINE, No. 104 State street, Albany, N. Y. *Local.*
673. Rev. JOHN A. PAINE, jr., Newark, N. J. *Local.*
674. CHARLES F. PARKER, Philadelphia, Pa. *Local.*
675. Prof. W. A. PARKER (Professor of ——, Iowa College), Grennele, Iowa. *Local.*
676. Dr. C. C. PARRY, Davenport, Iowa. *North American.* Special, *Rocky Mountains.*
677. Prof. T. L. PARVIN (Professor of Natural History, Iowa State University), Iowa City, Iowa. *Local.*
678. CHARLES H. PECK, Albany, N. Y. *Local.*
679. E. PECK, Washington, D. C. *North American.*
680. Prof. ROBERT PETER (Professor of Natural Science, Kentucky University), Lexington, Ky. *Local.*
681. THOMAS M. PETERS, Moulton, Ala. *Local.*
682. GEORGE D. PHIPPEN, Salem, Mass. *Local.* Special, *Wild Flowers under cultivation.*
683. Dr. ZINA PITCHER, Detroit, Mich. *North American.*
684. ISAAC A. POOL, No. 829 Washington street, Chicago, Ill. *Horticultural.*
685. B. S. PORTER, New Albany, Ind. *Local.*
686. Prof. THOMAS C. PORTER (Professor of Botany and Zoölogy, Lafayette College), Easton, Pa. *North American.* Special, *Pennsylvania.*
687. Prof. A. N. PRENTISS (Professor of Botany and Horticulture, Michigan State Agricultural College), Lansing, Mich. *North American.*
688. MANUEL J. PRESAS, Calle de Velarde, No. 5, Matanzas, Cuba. *Cuban.*
689. WM. H. RAND, Chicago, Ill. *Local.*
690. Dr. J. H. RAUCH, Chicago, Ill. *Local.*
691. H. W. RAVENEL, Aiken, S. C. *Local.* Special, *Fungi.*
692. THOMAS B. REDDING, Newcastle, Ind. *Local.*
693. Dr. SAMUEL REID, New Albany, Ind. *Local.*
694. JAMES RICHARDS, Litchfield, Ct. *Local.*
695. Dr. J. W. ROBBINS, Uxbridge, Mass. *North American.* Special, *Fresh Water Plants.*
696. JOSEPH T. ROTHROCK, McVeytown, Pa. *General.*
697. Prof. JOHN L. RUSSELL (Professor of Botany, Massachusetts Horticultural Society), Salem, Mass. *North American.* Special, *Cryptogamia.*

698-728 BOTANY. 32

698. Prof. ABRAM SAGER (Professor of ——, University of Michigan), Ann Arbor, Mich. *Local.*
699. Prof. J. H. SALISBURY (Professor of Physiology, Histology and Cell Pathology, Charity Hospital Medical College), Cleveland, Ohio. *Fungi.*
700. Dr. C. SARTORIUS, Mirader, Mexico. *Local.*
701. Dr. H. P. SARTWELL, Penn Yan, N. Y. *Local.* Special, *Carices.*
702. WILLIAM SAUNDERS, Dundas street, London, Canada West. *North American.*
703. FRANCISCO ADOLFO SAUVALLE, Habana, Cuba. *Cuban.*
704. GEORGE SCARBOROUGH, Sumner, Atchinson Co., Kansas. *Local.*
705. Prof. GEORGE C. SCHÆFFER, Washington, D. C. *North American.*
706. Dr. ARTHUR SCHOTT, Georgetown, D. C. *Mexico and Central America.*
707. R. ROBINSON SCOTT, Port Kennedy, Pa. *Local.*
708. THOMAS F. SEAL, Unionville, Chester Co., Pa. *Local.*
709. Miss LYDIA SHATTUCK, South Hadley, Mass. *Local.*
710. HENRY SHAW, St. Louis, Mo. *Local.*
711. JAMES M. SHAW, South Waterford, Me. *Local.*
712. Prof. D. S. SHELDON (Professor of Chemistry and Natural Sciences, Griswold College), Davenport, Iowa. *North American.*
713. Dr. A. G. SKINNER, Youngstown, Niagara Co., N. Y. *Local.*
714. AUBREY H. SMITH, No. 1516 Pine street, Philadelphia, Pa. *Local.*
715. CHARLES E. SMITH (President, Reading R. R. Co.), Philadelphia, Pa. *Local.*
716. DANIEL B. SMITH, Germantown, Pa. *Local.*
717. S. I. SMITH, New Haven, Ct. *New England.*
718. WM. R. SMITH (Superintendent United States Botanic Garden), Washington, D. C. *North American.*
719. C. J. SPRAGUE, Boston, Mass. *Fungi.*
720. ISAAC SPRAGUE, Grantville, Mass. *Botanical Artist.*
721. JACOB STAUFFER (Secretary, Linnæan Society of Lancaster), Lancaster, Pa. *Local.*
722. Rev. JAMES STEPHENSON, St. Inigos, St. Mary's Co., Md. *Local.*
723. Dr. GEORGE T. STEVENS, Albany, N. Y. *Local.*
724. Dr. —— STIVES, San Francisco, Cal. *Californian.* Special, *Algæ.*
725. SAMUEL STURTON, Quebec, Canada. *Canadian.*
726. WM. S. SULLIVANT, Columbus, Ohio. *General.* Special, *Musci.*
727. Rev. J. A. SWAN, Kennebunk, Me. *Local.*
728. EDWARD TATNALL, Wilmington, Del. *Local.*

729. Prof. Sanborn Tenney (Professor of Natural Sciences, Vassar Female College), Poughkeepsie, N. Y. *North American.*
730. Dr. John G. Thomas, Rivière-du-Loup-en-bas, Canada East. *Canadian.*
731. John J. Thomas, Union Springs, Cayuga Co., N. Y. *Local.*
732. Prof. George Thurber, Office, American Agriculturist, New York, N. Y. *North American.* Special, *Gramineæ.*
733. Prof. John Torrey (Professor of Botany, Columbia College), New York, N. Y. *General.* Special, *North American.*
734. Dr. Morton S. Townshend, Avon, Lorain Co., Ohio. *Local.*
735. C. M. Tracy (Curator of Botany, Essex Institute), Lynn, Mass. *General Collection.* Special, *New England.*
736. Prof. Edward Tuckerman (Professor of Botany, Amherst College), Amherst, Mass. *General.* Special, *Lichens.*
737. Dr. George Vasey, Richview, Washington Co., Ill. *North American.*
738. Wm. S. Vaux (Vice President and Curator, Academy of Natural Sciences), No. 1700 Arch street, Philadelphia, Pa. *North American.*
739. Prof. A. E. Verrill (Professor of Zoölogy, Yale College), New Haven, Ct. *New England.*
740. Dr. J. A. Warder, Cincinnati, Ohio. *Local.*
741. G. Warring, Boalsburg, Pa. *Local.*
742. David A. P. Watt (Editor, Canadian Naturalist and Geologist), Montreal, Canada. *Canadian.* Special, *Fungi.*
743. Miss Mary Whittington, Harrodsburg, Ky. *Local.*
744. Daniel Wilkins, Littleton, N. H. *Local.*
745. H. Willey, New Bedford, Mass. *Lichens.*
746. Prof. O. R. Willis, Whiteplains, N. Y. *North American.*
747. Hugh Wilson, Salem, Mass. *Horticultural.* Special, *Ferns under cultivation.*
748. Nathaniel Wilson (Curator, Island Botanic Garden), Jamaica, W. I. *West Indian.*
749. N. H. Winchell, Ann Arbor, Mich. *Local.*
750. I. R. Wirt, McVeytown, Pa. *Local.*
751. W. Wynne Wistar, Germantown, Pa. *Local.*
752. John Wolf, Canton, Fulton Co., Ill. *Local.*
753. Prof. Alphonso Wood, Brooklyn, N. Y. *North American.*
754. Prof. H. C. Wood, jr. (Professor of Botany, University of Pennsylvania), Philadelphia, Pa. *North American.*
755. Charles Wright, Wethersfield, Ct. *General.* Special, *Cuba, Texas, and New Mexico.*

## ARCHÆOLOGY.

756. T. A. CHENEY (Lib., Georgic Library), Havana, N. Y.  *N. Am.*
757. Dr. E. H. DAVIS, Worcester, Mass.  *American.*
758. Dr. SAMUEL A. GREENE, Boston, Mass.  *North American.*
759. SAMUEL F. HAVEN (Secretary, American Antiquarian Society), Worcester, Mass.  *North American.*
760. Hon. I. A. LAPHAM, Milwaukie, Wis.  *North American.*
761. Rev. SAMUEL LOCKWOOD, Keyport, N. J.  *New Jersey.*
762. Prof. O. C. MARSH (Professor of Palæontology, Yale College), New Haven, Ct.  *North American.*
763. Rev. ABNER MORSE, Boston, Mass.  *North American.*
764. FRANKLIN PEALE, No. 1131 Girard street, Philadelphia, Pa. *North American.*
765. CHARLES RAU, New York, N. Y.  *American.*
766. G. PEABODY RUSSELL (Curator of Archæology, Essex Institute), Salem, Mass.  *North American.*
767. E. GEORGE SQUIER, No. 105 East Thirty-ninth street, New York, N. Y.  *American.*
768. WM. S. VAUX (Vice President and Curator, Academy of Nat. Sciences), No. 1700 Arch street, Philadelphia, Pa.  *N. Am.*
769. Col. CHARLES WHITTLESEY, Cleveland, Ohio.  *North American.*
770. Dr. J. N. WILSON, Newark, Ohio.  *Local.*
771. Dr. A. WISLIZENUS, St. Louis, Mo.  *New Mexico, etc.*
772. Prof. JEFFRIES WYMAN (Professor of Comparative Anatomy and Physiology, Harvard University; President, Boston Society of Natural History), Cambridge, Mass.  *General.*

## ETHNOLOGY.

773. Rev. JOHN BACHMAN, Charleston, S. C.  *General.*
774. Dr. E. H. DAVIS, Worcester, Mass.  *American.*
775. GEORGE GIBBS, Washington, D. C.  *General.*
776. Dr. J. AITKEN MEIGS, Academy of Natural Sciences, Philadelphia, Pa.  *Craniology.*
777. LEWIS H. MORGAN, Rochester, N. Y.  *North American Indians.*
778. Dr. J. C. NOTT, Mobile, Alabama.  *General.*
779. Prof. HENRY S. PATTERSON, Philadelphia, Pa.  *General.*
780. Dr. CHARLES PICKERING, Boston, Mass.  *General.*
781. ALEX. S. TAYLOR, Santa Barbara, Cal.  *North American Indians.*
782. Prof. DANIEL WILSON (Professor of History and English Literature, University College), Toronto, C. W.  *General.*
783. Dr. A. WISLIZENUS, St. Louis, Mo.  *General.*

# GENERAL ZOÖLOGY.

784. Prof. Louis Agassiz (Professor of Zoölogy and Geology, Harvard University; Director and Curator, Museum of Comparative Zoölogy), Cambridge, Mass.
785. Prof. S. F. Baird (Assistant Secretary, Smithsonian Institution), Washington, D. C.
786. Prof. James D. Dana (Professor of Geology and Mineralogy, Yale College), New Haven, Ct.
787. Prof. J. W. Dawson (Principal, McGill University), Montreal, Canada.
788. Prof. Joseph Leidy (Professor of Anatomy, University of Pennsylvania; Curator, Academy of Natural Sciences of Philadelphia), No. 1302 Filbert street, Philadelphia, Pa.
789. Prof. Jeffries Wyman (Professor of Comparative Anatomy and Physiology, Harvard University; President, Boston Society of Natural History), Cambridge, Mass.

# MAMMALS.

790. Dr. Harrison Allen, Academy of Natural Sciences, Philadelphia, Pa. *Chiroptera.*
Rev. John Bachman, Charleston, S. C. *North American.*
791. Prof. S. F. Baird (Assistant Secretary, Smithsonian Institution), Washington, D. C. *American.*
792. George Barnston, Montreal, Canada. *Local.*
793. Dr. G. A. Canfield, Monterey, Cal. *Californian.*
794. Dr. J. G. Cooper, (Zoölogist, California State Survey; Curator of Zoölogy, California Academy of Natural Sciences), San Francisco, Cal. *Californian.*
795. Prof. E. D. Cope (Curator, Academy of Natural Sciences), Philadelphia, Pa. *Cetacea.*
796. Dr. Elliott Coues, U. S. A., Smithsonian Institution, Washington, D. C. *Arizonian.*
797. C. W. Gilbert (Assistant Curator of Mammalia, Worcester Society of Natural History), Worcester, Mass. *Local.*
798. Prof. Theo. Gill (Librarian, Smithsonian Institution), Washington, D. C. *General.*
799. Dr. John Gundlach, Calle de la Reina, 61, Habana, Cuba. *Cuban.*
800. Hon. Richard Hill, Spanishtown, Jamaica. *Jamaican.*

801. W. HUNTER (Taxidermist, Natural History Society of Montreal), Montreal, Canada. *Local.*
802. NATHANIEL PAINE (Curator of Mammalia, Worcester Society of Natural History), Worcester, Mass. *Local.*
803. TITIAN R. PEALE, Washington, D. C. *General.*
804. Prof. FELIPE POEY, Calle del Aguila, 157, Habana, Cuba. *Cuban.*
805. F. W. PUTNAM (Superintendent, Essex Institute; Curator of Ichthyology, Boston Society of Natural History), Salem, Mass. *Essex County, Mass.*
806. T. T. RICHARDS, St. Louis, Mo. *Crania.*
807. BERNARD R. ROSS, Rupert House. *Arctic.*
808. E. A. SAMUELS, Office State Board of Agriculture, Boston, Mass. *Local.*
809. Dr. J. H. SLACK, No. 1701 Spruce street, Philadelphia, Pa. *Quadrumana.*
810. Dr. GEORGE SUCKLEY, U. S. A., New York, N. Y. *Washington Territory.*
811. Prof. A. E. VERRILL (Professor of Zoölogy, Yale College; Curator of Radiata, Boston Society of Natural History, Boston, Mass.), New Haven, Ct. *North American.*
812. Dr. J. C. WHITE (Curator of Comparative Anatomy and Mammalogy, Boston Society of Natural History), Boston, Mass. *Anatomy.*
813. J. F. WHITEAVES (Curator and Rec. Secretary, Natural History Society of Montreal), Montreal, Canada. *Local.*

## BIRDS.*

814. JOHN AKHURST, No. 9½ Prospect street, Brooklyn, N. Y. *Local. Taxidermist and Dealer.*
815. J. A. ALLEN, Springfield, Mass. *New England.*
816. G. ALMA, Farmersville, Seneca Co., N. Y. *Local.*
817. Rev. JOHN AMBROSE, St. Margaret's Bay, Halifax Co., Nova Scotia. *Local.*
818. AMORY L. BABCOCK, Sherborn, Mass. *Surinam, S. A., and Local. Taxidermist.*
819. Prof. S. F. BAIRD (Assistant Secretary, Smithsonian Institution), Washington, D. C. *General.* Special, *American.*
820. VINCENT BARNARD, Kennett Square, Chester Co., Pa. *Local.*

* NOTE. Oölogy has now become so intimately connected with the study of the Birds themselves that about every person paying attention to Ornithology also has collections of the Eggs of Birds, therefore Oölogy, as a separate department, is omitted in the DIRECTORY.

821. GEORGE BARNSTON, Montreal, Canada. *Local.*
822. Rev. M. W. BEAUCHAMP, King's Ferry, Cayuga, Co., N. Y. *Local.*
823. S. B. BECKETT (Curator of Ornithology, Portland Society of Natural History), Portland, Me. *Local.*
824. JOHN G. BELL, No. 339 Broadway, New York, N. Y. *Taxidermist and Dealer.*
825. C. W. BENNETT (Curator of Ornithology, Museum of the Springfield City Library Association), Holyoke, Mass. *Local.*
826. Mrs. J. L. BODE, No. 16 North William street, New York, N. Y. *Taxidermist and Dealer.*
827. CHAS. L. BLOOD, Corner of Weir and First streets, Taunton, Mass. *Local. Taxidermist.*
828. G. A. BOARDMAN, Milltown, Me. *Local.*
829. S. H. BOWKER (Assistant Curator of Ornithology, Worcester Society of Natural History), Worcester, Mass. *Local.*
830. JOSEPH BRANO, Philadelphia, Pa. *Local. Taxidermist.*
831. Dr. T. M. BREWER (Curator of Oölogy, Boston Society of Natural History), Boston, Mass. *North American. Oölogy.*
832. C. G. BREWSTER, No. 16 Tremont street, Boston, Mass. *Dealer.*
833. E. A. BRIGHAM (Assistant Curator of Ornithology, Boston Society of Natural History), Boston, Mass. *Local.*
834. GEORGE C. BROWN (Curator and Treasurer, Burlington Co. Lyceum of Natural History), Mount Holly, N. J. *N. American.*
835. J. ELLIOTT CABOT (Curator of Ornithology, Boston Society of Natural History), Brookline, Mass. *General Collection.*
836. Dr. SAMUEL CABOT, Boston, Mass. *North American.*
837. R. A. CAMPBELL, Newark, Ohio. *Local.*
838. JOHN CASSIN (Vice President and Curator of Ornithology, Academy of Natural Sciences), Philadelphia, Pa. *General.*
839. RICHARD CHRIST, Nazareth, Pa. *Local.*
840. SAMUEL C. CLARK, Chicago, Ill. *North American.*
841. JOHN COLTON, Worcester, Mass. *Local.*
842. Dr. J. G. COOPER (Zoölogist, California State Survey; Curator of Zoölogy, California Academy of Natural Sciences), San Francisco, Cal. *Pacific Coast of N. A.*
843. THOMAS COTTLE, Woodstock, Canada West. *Local.*
844. Dr. ELLIOTT COUES, U. S. A., Smithsonian Institution, Washington, D. C. *American.*
845. WILLIAM COUPER (Vice President, Quebec Branch, Entomological Society of Canada), Quebec, Canada. *North American. Taxidermist.*
846. C. A. CRAIG, Montreal, Canada. *Local. Taxidermist.*

847. Dr. JOHN DARBY, South Williamstown, Mass. *Local.*
848. HENRY DAVIS, McGregor, Iowa. *Local.*
849. J. C. DEACON, Chicopee, Mass. *Local. Taxidermist.*
850. RAFAEL MONTES DE OCA, Xalapa, Mexico. *Mexican.*
851. C. DREXLER, Washington, D. C. *Local. Taxidermist.*
852. D. G. ELLIOT, No. 27 West Thirty-third street, New York, N. Y. *American.*
853. SAMUEL E. ELMORE, Hartford, Ct. *Local.*
854. W. E. ENDICOTT, Canton, Mass. *New England.*
855. Prof. H. FAIRBANKS (Prof. of Natural Philosophy, Dartmouth College), Hanover, N. H. *North American.*
856. CHARLES FELDMAN, Philadelphia, Pa. *Local. Taxidermist.*
857. WM. H. FLOYD, Weston, Mass. *North American.*
858. AUGUSTUS FOWLER, Danvers, Mass. *Local.*
859. SAMUEL P. FOWLER (Vice President, Essex Institute, Salem), Danvers, Mass. *Local.*
860. Dr. A. VON FRANTZIUS, San José, Costa Rica. *Costa Rican.*
861. ALEXANDER GALBRAISH, No. 209 North Ninth street, Philadelphia, Pa. *Local. Taxidermist and Dealer.*
862. CHARLES GALBRAITH, West Hoboken, N. J. *Local. Taxidermist.*
863. WM. GALBRAITH, West Hoboken, N. J. *Local. Taxidermist.*
864. WM. L. GILL, Lancaster, Pa. *Local.*
865. Col. A. J. GRAYSON, Mazatlan, Mexico. *Mexican.*
866. FERD. GRUBER, San Francisco, Cal. *Local. Taxidermist.*
867. Dr. JOHN GUNDLACH, Calle de la Reina, 61, Habana, Cuba. *West Indian.*
868. Prof. C. E. HAMLIN (Professor of Natural History, Waterville College), Waterville, Me. *Maine.*
869. HENRY HANFORD, Columbus, Ohio. *Local.*
870. GEORGE HENSEL, Lancaster, Pa. *Local. Taxidermist.*
871. JAMES HEPBURN, San Francisco, Cal. *Western Coast of America.*
872. Dr. A. HALL, Montreal, Canada. *Local.*
873. Hon. RICHARD HILL, Spanishtown, Jamaica. *Local.*
874. THOMAS HOLE, Clarkson, Columbiana Co., Ohio. *Local.*
875. Dr. P. R. HOY, Racine, Wis. *Local.*
876. Dr. A. T. HUDSON, U. S. A., Lyons, Clinton Co., Iowa. *Local.*
877. D. DARWIN HUGHES, Marshall, Mich. *Local.*
878. CHAS. A. HOUGHTON, Holliston, Mass. *Local. Taxidermist.*
879. W. HUNTER (Taxidermist, Natural History Society of Montreal), Montreal, Canada. *Local.*
880. ILGES and SANTER, No. 15 Frankfort street, New York, N. Y. *Taxidermists.*

881.  JOHN JENKINS, Monroe, Orange Co., N. Y. *Local. Taxidermist.*
882.  SAMUEL JILLSON, Hudson, Mass. *Local. Taxidermist.*
883.  Rev. C. M. JONES, North Madison, Ct. *Local.*
884.  F. KÆMPFER, Chicago, Ill. *Local.*
885.  JOHN KIRKPATRICK (Secretary, Academy of Natural Sciences of Cleveland; Secretary, Cleveland Horticultural Society), Cleveland, Ohio. *Local.*
886.  Rev. A. B. KENDIG, Davenport, Iowa. *Local.*
887.  Dr. JARED P. KIRTLAND, East Rockport, Ohio, P. O. address, Cleveland, Ohio. *Local.*
888.  Miss H. M. KNOWLTON (Assistant Curator of Ornithology, Worcester Society of Nat. History), Worcester, Mass. *Local.*
889.  JOHN KRIDER, Corner of Second and Walnut streets, Philadelphia, Pa. *North American. Taxidermist.*
890.  THURE KUMLIEN, Busseyville P. O., Wis. *North American.*
891.  GEORGE N. LAWRENCE, No. 172 Pearl street, New York, N. Y. *American.*
892.  J. S. LEACH, Bridgewater, Mass. *Local.*
893.  JAMES M. LEMOINE, Quebec, Canada. *Local.*
894.  Miss F. S. LINCOLN (Curator of Oölogy, Worcester Society of Natural History), Worcester, Mass. *Local Oölogy.*
895.  Dr. STARLING LORING, Columbus, Ohio. *Local.*
896.  B. P. MANN, Cambridge, Mass. *Local Oölogy.*
897.  WILLIAM T. MARCH, Spanishtown, W. I. *West Indian.*
898.  Dr. ISAAC C. MARTINDALE (Director, Byberry Philosophical Society), Byberry, Pa. *Local.*
899.  L. J. MAYNARD, Newtonville, Mass. *Local. Taxidermist.*
900.  R. McFARLANE, Fort Anderson, British America. *Northern Regions of America.*
901.  THOMAS McILWRAITH, Hamilton, C. W. *Local.*
902.  Prof. MANLY MILES (Professor of Animal Physiology and Practical Agriculture, State Agricultural College), Lansing, Mich.
903.  CHARLES H. NAUMAN (Chairman, Committee on Ornithology, Linnæan Society of Lancaster), Box 508, Lancaster, Pa. *Local.*
904.  Dr. J. S. NEWBERRY (Professor of Geology, Columbia College), New York, N. Y. *North American.*
905.  GEORGE Y. NICKERSON, No. 42 Williams street, New Bedford, Mass. *Local. Taxidermist.*
906.  J. P. NORRIS, Philadelphia, Pa. *Local.*
907.  CHARLES S. PAINE, East Bethel, Vt. *Local.*
908.  FREDERICK PASSMORE, Yonge street, Toronto, C. W. *North American.*

909. TITIAN R. PEALE, Washington, D. C. *General.*
910. Dr. D. WEBSTER PRENTISS, Washington, D. C. *Local.*
911. HENRY A. PURDIE, Boston, Mass. *Local.*
912. F. W. PUTNAM (Superintendent, Essex Institute; Curator of Ichthyology, Boston Society of Natural History), Salem, Mass. *Essex County, Mass.*
913. L. E. RICKSECKER, Nazareth, Pa. *Local.*
914. O. RIENECKE (Curator of Ornithology, Buffalo Society of Natural Sciences), Buffalo, N. Y. *Local.*
915. A. H. RIISE, St. Thomas, W. I. *West Indian.*
916. JAMES S. ROGERS, New York, N. Y. *Local.*
917. JAMES H. ROOME, No. 55 Carmine street, New York, N. Y. *Local. Taxidermist.*
918. BERNARD R. ROSS, Rupert House. *Arctic.*
919. WILLIAM A. ROUSSEAU, Troy, N. Y. *Local.*
920. E. A. SAMUELS, Office State Board of Agriculture, Boston, Mass. *New England. North American Oölogy.*
921. Dr. C. SARTORIUS, Mirador, Mexico. *Mexican.*
922. LIVINGSTON SATTERLEE, New York, N. Y. *Local.*
923. JOHN H. SEARS, Danvers, Mass. *Essex County, Mass.*
924. HENRY SHAW (Assistant Curator of Ornithology, Worcester Society of Natural History), Worcester, Mass. *Local.*
925. —— SHERMAN, Bedford street, New Bedford, Mass. *Local.*
926. Prof. HENRY SHIMER, Mt. Carroll, Ill. *Local.*
927. JAMES G. SHUTE, Woburn, Mass. *Local.*
928. S. I. SMITH, New Haven, Ct. *New England.*
929. WILLIAM A. SMITH (Curator of Ornithology, Worcester Society of Natural History), Worcester, Mass. *Local.*
930. Rev. WM. S. SOUTHGATE, Litchfield, Ct. *Local.*
931. JACOB STAUFFER (Secretary, Linnæan Society of Lancaster), Lancaster, Pa. *Local.*
932. Rev. JAMES STEPHENSON, St. Inigos, St. Mary's Co., Md. *Local.*
933. Dr. GEORGE SUCKLEY, U. S. A., New York, N. Y. *North American.*
934. JOSEPH SULLIVANT, Columbus, Ohio. *Local.*
935. Prof. F. SUMICHRAST, Orozaba, Mexico. *Mexican.*
936. E. L. SUMNER (Assistant Curator of Oölogy, Worcester Society of Natural History), Worcester, Mass. *Local.*
937. S. H. SYLVESTER, Middleborough, Mass. *Local. Taxidermist.*
938. JAMES TAYLOR, Philadelphia, Pa. *Local. Taxidermist.*
939. T. MARTIN TRIPPE, Orange, N. Y. *Local.*
940. H. G. VERNOR, Montreal, Canada. *Local.*

941. Dr. VELIE, Bath, Steuben Co., N. Y. *North American.*
942. Prof. A. E. VERRILL (Professor of Zoölogy, Yale College; Curator of Radiata, Boston Society of Natural History), New Haven, Ct. *North American.*
943. NATHANIEL VICKARY, No. 262 Chestnut street, Lynn, Mass. *Local. Taxidermist and Dealer.*
944. FREDERIC WARE, Cambridge, Mass. *Local.*
945. JOHN M. WHEATON, Columbus, Ohio. *Local.*
946. J. F. WHITEAVES (Recording Secretary and Curator, Natural History Society of Montreal), Montreal, Canada. *Canadian.*
947. Prof. WM. D. WHITNEY (Professor of Sanskrit, Yale College), New Haven, Ct. *New England.*
948. ROBERT WILSON, Gouverneur, N. Y. *Local.*
949. R. K. WINSLOW, Cleveland, Ohio. *Local.*
950. ALEXANDER WOLLE, Baltimore, Md. *Local. Taxidermist.*
951. C. J. WOOD, Philadelphia, Pa. *Local. Taxidermist.*
952. Dr. WM. WOOD, East Windsor Hill, Ct. *Local.*

## REPTILES.

953. Prof. LOUIS AGASSIZ (Professor of Zoölogy and Geology, Harvard University; Director and Curator, Museum of Comparative Zoölogy), Cambridge, Mass. *General. Special, Chelonia.*
954. Prof. S. F. BAIRD (Assistant Secretary, Smithsonian Institution), Washington, D. C. *General Collection. Special, N. American.*
955. Prof. EDWARD D. COPE (Curator, Academy of Natural Sciences of Philadelphia), Philadelphia, Pa. *General.*
956. Dr. B. F. FOGG (Curator of Herpetology, Portland Society of Natural History), Portland, Me. *Local.*
957. Prof. J. E. HOLBROOK, Charleston, S. C. *North American.*
958. Prof. JOHN LE CONTE (Professor of Natural Philosophy, University of South Carolina), Columbia, S. C. *Local.*
959. Prof. JOSEPH LEIDY (Professor of Anatomy, University of Pennsylvania; Curator, Academy of Natural Sciences), No. 1302 Filbert street, Philadelphia, Pa. *Fossil.*
960. Prof. O. C. MARSH (Professor of Palæontology, Yale College), New Haven, Ct. *Fossil.*
961. Prof. J. S. NEWBERRY (Professor of Geology, Columbia College), New York, N. Y. *Fossil.*
962. F. W. PUTNAM (Superintendent, Essex Institute; Curator of Ichthyology, Boston Society of Natural History; Editor, American Naturalist), Salem, Mass. *General Collection. Special, North American.*

963. STEPHEN SALISBURY, jr. (Curator of Herpetology, Worcester Society of Natural History), Worcester, Mass. *Local.*
964. JACOB STAUFFER (Secretary, Linnæan Society of Lancaster), Lancaster, Pa. *Local.*
965. Prof. A. E. VERRILL (Professor of Zoölogy, Yale College; Curator of Radiata, Boston Society of Natural History), New Haven, Ct. *North American.*
966. Dr. B. G. WILDER (Professor of Natural History, Cornell University; Curator of Herpetology, Boston Society of Natural History; Assistant, Museum of Comp. Zoölogy), Boston, Mass. *General Collection.*
967. HENRY S. WILLIAMS, Ithaca, N. Y. *Local.*
968. Prof. ALEXANDER WINCHELL (Professor of Natural History, University of Michigan), Ann Arbor, Mich. *Local.*
969. Dr. T. G. WORMLEY, Columbus, Ohio. *Local.*

## FISHES.

970. Dr. C. C. ABBOTT (Zoölogist, New Jersey State Survey), Trenton, N. J. *Local.*
971. ALEX. E. R. AGASSIZ (Assistant, Museum of Comparative Zoölogy), Cambridge, Mass. *Embiotocoids.*
972. Prof. LOUIS AGASSIZ (Professor of Zoölogy and Geology, Harvard University; Curator and Director, Museum of Comparative Zoölogy), Cambridge, Mass. *General.*
973. G. ALMA, Farmersville, Seneca Co., N. Y. *Local.*
974. Rev. JOHN AMBROSE, St. Margaret's Bay, Halifax Co., Nova Scotia. *Local.*
975. CHARLES G. ATKINS (Commissioner of River Fisheries), Augusta, Me. *Fishculture.*
976. Capt. N. E. ATWOOD, Provincetown, Mass. *Local. Food Fishes.*
977. Dr. W. O. AYRES (Corresponding Secretary, California Academy of Natural Sciences), San Francisco, Cal. *Californian.*
978. Prof. S. F. BAIRD (Assistant Secretary, Smithsonian Institution), Washington, D. C. *North American.*
979. Rev. M. W. BEAUCHAMP, King's Ferry, Cayuga, N. Y. *Local.*
980. J. CARSON BREVOORT (President, Long Island Historical Society), Brooklyn, N. Y. *General.*
981. Dr. ROBERT BRIDGES, Philadelphia, Pa. *Local.*
982. SAMUEL C. CLARK, Chicago, Ill. *Local.*
983. J. D. COOPER, San Francisco, Cal. *Californian.*
984. Prof. EDWARD D. COPE (Curator, Academy of Natural Sciences of Philadelphia), Philadelphia, Pa. *General Collection.* Special, *North American Fresh-water.*

985. Dr. J. W. Dawson (Principal, McGill University), Montreal, Canada. *Local and Fossil.*

986. Andrew Garrett, Care of Samuel Hubbard, Agent Pacific Mail Steamship Co., San Francisco, Cal. *South Seas.*

987. Dr. W. P. Gibbons, Alameda Co., Cal. *Embiotocoids.*

988. Prof. Theodore Gill (Librarian, Smithsonian Institution), Washington, D. C. *General.* .

989. Dr. J. Bernard Gilpin (Vice President, Nova Scotian Institute of Natural Science), Halifax, Nova Scotia. *Nova Scotian.*

990. Dr. A. C. Hamlin, Bangor, Me. *Salmonidæ of Maine.*

991. Hon. Richard Hill, Spanishtown, Jamaica. *West Indian.*

992. Prof. J. E. Holbrook, Charleston, S. C. *Southern States.*

993. J. Matthew Jones, Ashbourne, Nova Scotia. *Nova Scotian.*

994. Dr. J. P. Kirtland, East Rockport, Ohio; Post-office address, Cleveland, Ohio. *Great Lakes and Ohio.*

995. Prof. Joseph Leidy (Professor of Anatomy, University of Pennsylvania; Curator, Academy of Natural Sciences), No. 1302 Filbert street, Philadelphia, Pa. *Fossil.*

996. Rev. Samuel Lockwood, Keyport, N. J. *Local.*

997. Theodore Lyman (Assistant, Museum of Comp. Zoölogy, Cambridge; Commissioner of River Fisheries), Brookline, Mass. *Fishculture.*

998. Dr. R. P. Mann, Milford, Ohio. *Devonian.*

999. H. A. Marsh (Assistant Curator of Ichthyology, Worcester Society of Natural History), Worcester, Mass. *Local.*

1000. Prof. O. C. Marsh (Professor of Palæontology, Yale College), New Haven, Ct. *Fossil.*

1001. Dr. J. C. Morris, Philadelphia, Pa. *Local.*

1002. Dr. William A. Nason, Post-office box 3412, Chicago, Ill. *Local.*

1003. Prof. J. S. Newberry (Professor of Geology, Columbia College), New York, N. Y. *Fossil.*

1004. Thaddeus Norris, 505 Minor st., Philadelphia, Pa. *Game Fishes.*

1005. Prof. Felipe Poey, Calle del Aguila, 157, Habana, Cuba. *West*

1006. M. N. Preston, Skaneateles, N. Y. *Local.*          [*Indian.*

1007. F. W. Putnam (Superintendent, Essex Institute; Curator of Ichthyology, Boston Society of Natural History; Editor, American Naturalist), Salem, Mass. *General.*

1008. O. H. St. John, Waterloo, Iowa. *Fossil.*

1009. S. H. Scudder (Custodian, Secretary, Librarian, and Curator of Entomology, Boston Society of Natural History), Cambridge, Mass. *Hæmulidæ.*

1010. H. L. Shumway (Assistant Curator of Ichthyology, Worcester Society of Natural History), Worcester, Mass. *Local.*

1011. Dr. D. H. STORER, Boston, Mass. *Massachusetts.*
1012. Dr. GEORGE SUCKLEY, U. S. A., New York, N. Y. *Salmonidæ.*
1013. Dr. B. G. WILDER (Professor of Natural History, Cornell University; Assistant, Museum of Comparative Zoölogy; Curator of Herpetology, Boston Society of Natural History), Boston, Mass. *Selachians.*
1014. Prof. ALEXANDER WINCHELL (Professor of Natural History, University of Michigan), Ann Arbor, Mich. *North American and Fossil.*
1015. J. W. YOUNG, Cleveland, Ohio. *Local.*

## INSECTS.

1016. CHARLES E. AARON, Mount Holly, N. J. *Local.*
1017. ALVEY A. ADEE, No. 54 Exchange Place, New York, N. Y. *Hymenoptera.*   .
1018. A. E. R. AGASSIZ (Assistant, Museum of Comparative Zoölogy), Cambridge, Mass. *General.* Special, *Lepidoptera.*
1019. Prof. LOUIS AGASSIZ (Professor of Zoölogy and Geology, Harvard University; Curator and Director, Museum of Comparative Zoölogy), Cambridge, Mass. *General.*
1020. JOHN AKHURST, No. 9½ Prospect street, Brooklyn, N. Y. *Coleoptera.*
1021. W. P. ALCOTT, Andover, Mass. *Local.*
1022. J. A. ALLEN, Springfield, Mass. *Local.*
1023. JAMES ANGUS, West Farms, N. Y. *Hymenoptera and Lepidoptera.*
1024. T. B. ASHTON, North White Creek, Washington Co., N. Y. *Local.*
1025. E. P. AUSTIN, Cambridge, Mass. *Coleoptera.*
1026. SAMUEL AUXER, Lancaster, Pa. *Local.*
1027. AMORY L. BABCOCK, Sherborn, Mass. *Local, and Surinam, S. A.*
1028. AUSTIN BACON, Natick, Mass. *Local.*
1029. VINCENT BARNARD, Kennett Square, Chester Co., Pa. *Local.*
1030. HOMER F. BASSETT, Waterbury, Ct. *N. American Hymenoptera.*
1031. D. W. BEADLE, St. Catharines, Canada West. *Local Coleoptera and Lepidoptera.*
1032. Dr. H. BEHR (Entomologist, California State Board of Agriculture), San Francisco, Cal. *Californian.*
1033. JAMES S. BEHRENS, San Francisco, Cal. *Coleoptera and Lepidoptera.*
1034. G. W. BELFRAGE, Chicago, Ill. *A Collector.*
1035. AARON B. BELKNAP, New York, N. Y. *Local.*
1036. J. F. BENNER, New Lisbon, Ohio. *Local.*
1037. C. W. BENNETT, Holyoke, Mass. *Local.*

1038. Rev. CHARLES J. S. BETHUNE (Secretary and Treasurer, Entomological Society of Canada), Credit, Canada West. *Canadian Coleoptera, and North American Lepidoptera.*
1039. B. BILLINGS, Ottawa City, Canada West. *Local.*
1040. E. BILLINGS. Montreal, Canada. *Coleoptera.*
1041. CHARLES A. BLAKE, Philadelphia, Pa. *Local Lepidoptera.*
1042. M. P. BLAKE, Gilmanton, N. H. *Local.*
1043. J. H. B. BLAND, Philadelphia, Pa. *North American Coleoptera.*
1044. Col. J. H. BLISS, Erie, Pa. *Local.*
1045. CHARLES L. BLOOD, Corner of Weir and First streets, Taunton, Mass. *Local.*
1046. ANDREW BOLTER, Corner of Wells and Van Buren streets, Chicago, Ill. *Lepidoptera.*
1047. JOHN BOLTON, Portsmouth, Ohio. *Local.*
1048. —— BOTTIN, Orizaba, Mexico. *Mexican.*
1049. GEORGE J. BOWLES (Secretary, Quebec Branch, Entomological Society of Canada; Curator, Literary and Historical Society of Quebec), Quebec, Canada. *North American Lepidoptera.*
1050. FREDERICK BRACHES, Gray's Summit, Franklin Co., Mo. *Local.*
1051. GEORGE E. BRACKETT, Belfast, Me. *Local.*
1052. Rev. J. H. BRAKELEY, Bordentown, N. J. *Local.*
1053. Rev. W. B. BREED, Philadelphia, Pa. *Local.*
1054. Dr. EMIL BRENDEL, U. S. A., Peoria, Ill. *Pselaphids.*
1055. JOSEPH BRIDGHAM, jr., No. 26 Waverly Place, New York, N. Y. *North American Lepidoptera.*
1056. Mrs. JOSEPH BRIDGHAM, No. 26 Waverly Place, New York, N. Y. *Local.*
1057. ROBERT H. BROWNE, Quebec, Canada. *N. American Lepidoptera,*
1058. S. B. BUCKLEY, Geological Bureau, Austin, Texas. *Local. Formicidæ.*
1059. ROBERT BUNKER, Rochester, N. Y. *Local.*
1060. EDWARD BURGESS, Boston, Mass. *Local.*
1061. STEPHEN CALVERLY, Brooklyn, N. Y. *Local.*
1062. WILLIAM W. CAREY, Colerain, Mass. *Apiarian.*
1063. DAVID A. CASHMAN, Chicago, Ill. *Local.*
1064. H. H. CHAPMAN, Chicago, Ill. *Local.*
1065. JOSEPH E. CHASE, Holyoke, Mass. *Local.*
1066. F. A. CLAPP, Dorchester, Mass. *Local.*
1067. Rev. V. CLEMENTI, North Douro, Canada West. *Local.*
1068. RICHARD COLVIN, Baltimore, Md. *Apiarian.*
1069. CALEB COOKE (Curator of Articulata, Essex Institute), Salem, Mass. *Local. General Collection.*

1070. WILLIAM COUPER (Vice President, Quebec Branch, Entomological Society of Canada), Quebec, Canada. *Coleoptera and Insect Architecture.*

1071. E. T. CRESSON (Corresponding Secretary and Curator, American Entomological Society), No. 518 South Thirteenth street, Philadelphia, Pa. *General.* Special, *Hymenoptera.*

1072. Prof. HENRY CROFT (Professor of Chemistry, University College; President, Entomological Society of Canada), Toronto, Canada West. *North American Coleoptera.*

1073. CHARLES CURRIER, Providence, R. I. *Local.*

1074. W. O. CURRIER, Providence, R. I. *Local.*

1075. HENRY DAVIS, McGregor, Iowa. *Local.*

1076. A. W. DE FOREST, New York, N. Y. *North American.*

1077. THOMAS A. DICKINSON, Worcester, Mass. *Local.*

1078. GEORGE B. DIXON, (Librarian, American Entomological Society), Philadelphia, Pa. *General.*

1079. CHARLES R. DODGE (Assistant in Entomology, U. S. Department of Agriculture), Washington, D. C. *Local.*

1080. Dr. EDWARD DORSCH, Monroe, Mich. *Local.*

1081. WILLIAM H. EDWARDS, Newburgh, N. Y. *North American Lepidoptera.*

1082. JAMES H. EMERTON (Curator of Articulata, Essex Institute), Salem, Mass. *Local.* Special, *Arachnides. General Collection. Zoölogical Artist.*

1083. CHARLES A. EMERY, Springfield, Mass. *Local.*

1084. L. ENGELBRECHT, Portsmouth, Sciota Co., Ohio. *North American Coleoptera, Neuroptera, Lepidoptera.*

1085. J. M. ENGLISH, Chicago, Ill. *Local.*

1086. FRANK FAIRBANKS, St. Johnsbury, Vt. *Local.*

1087. CHARLES E. FAXON, Jamaica Plain, Mass. *Local.*

1088. WALTER FAXON, Jamaica Plain, Mass. *Local.*

1089. H. TUDOR FAY, Columbus, Ohio. *North American Coleoptera.*

1090. HENRY FELDMAN, Philadelphia, Pa. *Local Coleoptera.*

1091. WILLIAM C. FISH, East Falmouth, Mass. *Local.*

1092. Judge FISHBANK, Batavia, N. Y. *Apiarian.*

1093. D. M. FISK, Brown University, Providence, R. I. *Local.*

1094. Dr. ASA FITCH (State Entomologist of New York), Salem, Washington Co., N. Y. *General.* Special, *Injurious and Beneficial.*

1095. C. FOLEY, Lindsay, Canada West. *Local.*

1096. R. J. FOWLER, Montreal, Canada. *Lepidoptera.*

1097. ROBERT FRAZER, Philadelphia, Pa. *North American Coleoptera.*

1098. ALBERT S. GARLAND, Gloucester, Mass. *Local.*

1099. Dr. G. P. GIRDWOOD, Montreal, Canada. *Local.*

1100. TOWNEND GLOVER (Entomologist, U. S. Department of Agriculture), Washington, D. C. *North American.*
1101. EDWARD L. GRAEF (Member of Committee on Entomology, Long Island Historical Society), Brooklyn, N. Y. *North American and European Lepidoptera.*
1102. Dr. JOHN W. GREENE (Curator, Lyceum of Natural History of New York), No. 7 West Fifteenth street, New York, N. Y. *North American Hymenoptera.*
1103. AUGUSTUS R. GROTE (Curator of Entomology, Buffalo Society of Natural Science), No. 41 Beaver street, New York, N. Y. *North American Lepidoptera.*
1104. FERDINAND GRUBER, San Francisco, Cal. *Local.*
1105. Miss C. GUILD, Walpole, Mass. *Local.*
1106. Dr. JUAN GUNDLACH, Habana, Cuba. *Cuban.*
1107. Dr. HERMANN HAGEN (Assistant, Museum of Comparative Zoölogy), Cambridge, Mass. *General.* Special, *Neuroptera.*
1108. REUBEN HAINES, Germantown, Pa. *Lepidoptera.*
1109. Prof. S. S. HALDEMAN, Columbia, Pa. *N. American Coleoptera.*
1110. Dr. A. HALL, Montreal, Canada. *Local.*
1111. GEORGE H. HATHEWAY, Post-office box 5868, Chicago, Ill. *North American Hemiptera and Neuroptera.*
1112. Dr. G. W. HAZLETINE, Jamestown, N. Y. *Local.*
1113. Dr. CHARLES A. HELMUTH, Chicago, Ill. *N. American Coleoptera.*
1114. FRANK F. HODGMAN, Littleton, N. H. *Local.*
1115. GUSTAVE P. HOFFMAN, Chicago, Ill. *Local.*
1116. Dr. GEORGE H. HORN (President, American Entomological Socety), Academy of Natural Sciences, Philadelphia, Pa. *North American Coleoptera.*
1117. WINSLOW J. HOWARD, No. 345 Grand street, New York, N. Y. *Local.*
1118. ROBERT HOWELL, Nichols, Tioga Co., N. Y. *Local.*
1119. CHARLES N. HOYT, Providence, R. I. *Local.*
1120. Rev. JAMES HUBBERT (Professor of Natural Sciences, St. Francis College), Richmond, Canada East. *Diptera.*
1121. GEORGE HUNT, Providence, R. I. *North American.*
1122. Miss M. E. HUNT, Providence, R. I. *Local.*
1123. Miss M. L. JENKS (Assistant Curator of Articulata, Worcester Society of Natural History), Worcester, Mass. *Local.*
1124. Rev. W. A. JOHNSON, Weston, Canada West. *Local.*
1125. J. MATTHEW JONES, Ashbourne, Nova Scotia. *Local.*
1126. Prof. SAMUEL JONES (Professor of Natural Sciences, Jefferson College), Canonsburg, Pa. *Local.*
1127. WILLIAM KEOSTLIN, New York, N. Y. *Coleoptera.*

1128. Rev. P. P. KIDDER, Ellicottsville, N. Y. *Local.*
1129. JOHN KIRKPATRICK (Secretary, Academy of Natural Sciences of Cleveland; Secretary, Cleveland Horticultural Society), Cleveland, Ohio. *Local.*
1130. Dr. J. P. KIRTLAND, East Rockport, Ohio; Post-office address, Cleveland, Ohio. *Local Lepidoptera.*
1131. J. FRANK KNIGHT (Recording Secretary, American Entomological Society), Philadelphia, Pa. *North American Homoptera.*
1132. EDWARD KOCH, Toledo, Ohio. *Local.*
1133. Rev. L. L. LANGSTROTH, Oxford, Butler Co., Ohio. *Apiarian.*
1134. Dr. WILLIAM LE BARON, Geneva, Kane Co., Ill. *North American Diptera and Coleoptera.*
1135. Dr. JOHN L. LE CONTE, No. 1325 Spruce street, Philadelphia, Pa. *Coleoptera.*
1136. Dr. JAMES LEWIS, Mohawk, N. Y. *Local.*
1137. Dr. SAMUEL LEWIS, No. 1330 Spruce street, Philadelphia, Pa. *North American Coleoptera.*
1138. Dr. G. LINCECUM, Long Point, Texas. *Local.*
1139. J. A. LINTNER, Utica, N. Y. *Lepidoptera.*
1140. Miss ADELIA J. LITTLEFIELD, Woburn, Mass. *Local.*
1141. B. P. MANN, Cambridge, Mass. *Local.*
1142. Prof. R. Z. MASON, Appleton, Wis. *Local.*
1143. JAMES W. MCALLISTER (Treasurer, American Entomological Society), Philadelphia, Pa. *Hymenoptera.*
1144. THEODORE L. MEAD, No. 233 West Thirty-fourth street, New York, N. Y. *Local.*
1145. F. E. MELSHEIMER, Davidsburg, York Co., Pa. *General.*
1146. JOHN MEICHEL, Philadelphia, Pa. *Local.*
1147. JAMES C. MERRILL, Pemberton Square, Boston, Mass. *Local.*
1148. JULIUS E. MEYER, Brooklyn, N. Y. *Lepidoptera.*
1149. Prof. MANLY MILES (Professor of Animal Physiology and Practical Agriculture, State Agricultural College), Lansing, Mich. *Local.*
1150. Rev. J. G. MORRIS, Baltimore, Md. *North American Lepidoptera.*
1151. Dr. WILLIAM A. NASON, Post-office Box 3412, Chicago, Ill. *Local Coleoptera.*
1152. JAMES NEAL, Cleveland, Ohio. *Local.*
1153. GEORGE W. NICHOLS, West Amesbury, Mass. *Local.*
1154. J. NIETS, Cordova, Mexico. *Local.*
1155. D. BENJAMIN NORRIS, Pittsfield, Pike Co., Ill. *Local.*
1156. EDWARD NORTON, Farmington, Ct. *General.* Special, *North American Hymenoptera.*
1157. JOHN ORNE, jr., Cambridgeport, Mass. *N. American Coleoptera.*

1158. JOHN OSGOOD, Lynn, Mass. *Local.*
1159. Baron R. VON OSTEN SACKEN (Russian Consul General), No. 52 Exchange Place, New York, N. Y. *Diptera. Cynipidæ.*
1160. Dr. A. S. PACKARD, jr. (Curator of Crustacea, Boston Society of Natural History; Curator of Articulata, Essex Institute; Editor, American Naturalist), Salem, Mass. *General.* Special, *Hymenoptera and Lepidoptera.*
1161. TITIAN R. PEALE, Washington, D. C. *Lepidoptera.*
1162. CHARLES H. PECK, Albany, N. Y. *Local.*
1163. GEORGE WM. PECK, No. 26 Dey street, New York, N. Y. *North American Lepidoptera.*
1164. JOHNSON PETTIT, Grimsby, Canada West. *Local Coleoptera and Lepidoptera.*
1165. WM. S. PINE (Vice President, American Entomological Society), Philadelphia, Pa. *Local.*
1166. JAMES H. POE, Portsmouth, Ohio. *Local.*
1167. Prof. FELIPE POEY, Calle del Aguila, No. 157, Habana, Cuba. *Cuban Neuroptera and Formicidæ.*
1168. S. S. RATHVON, Lancaster, Pa. *Coleoptera.*
1169. TRYON REAKIRT, No. 353 North Third street, Philadelphia, Pa. *Diurnal Lepidoptera.*
1170. A. S. REBER, Bellefonte, Pa. *Local.*
1171. E. BAYNES REED (Secretary, London Branch, Entomological Society of Canada), London, C. W. *Local.*
1172. T. REYNOLDS, Montreal, Canada. *Local.*
1173. HARVEY J. RICH, Brooklyn, N. Y. *Local.*
1174. WM. J. RICHARDSON, Oxford, Miss. *Local.*
1175. JAMES RIDINGS, Philadelphia, Pa. *General.*
1176. JAMES H. RIDINGS, Philadelphia, Pa. *North American Neuroptera and Orthoptera.*
1177. A. H. RIISE, St. Thomas, West Indies. *West Indian.*
1178. C. V. RILEY, Chicago, Ill. *Local. Injurious to Vegetation.*
1179. A. S. RITCHIE, Montreal, Canada. *Local.*
1180. COLEMAN T. ROBINSON, New York, N. Y. *N. Am. Lepidoptera.*
1181. Dr. G. O. ROGERS, Lancaster, N. H. *Local.*
1182. WILLIAM A. ROUSSEAU, Troy, N. Y. *Local.*
1183. ALEX. L. RUSSELL, Quebec, Canada. *North American Lepidoptera.*
1184. J. SACHS, West Hoboken, N. J. *Local.*
1185. Prof. A. SAGER (Professor of Obstetrics, University of Michigan), Ann Arbor, Mich. *Local.*
1186. JAMES SAMPSON, New Harmony, Ind. *Local.*
1187. F. G. SANBORN, Boston Society of Natural History, Boston, Mass. *North American.*

1188. Dr. E. SANGER, Littleton, N. H.  *Local.*
1189. Dr. JOHN H. SANGSTER, Normal School, Toronto, C. W.  *Local.*
1190. Dr. C. SARTORIUS, Mirador, Mexico.  *Mexican.*
1191. WILLIAM SAUNDERS, Dundas street, London, Canada West. *North American Coleoptera and Lepidoptera.*
1192. MAURICE SHUSTER, St. Louis, Mo.  *North American Coleoptera.*
1193. S. H. SCUDDER (Custodian, Secretary, Librarian, and Curator of Entomology, Boston Society of Natural History), Cambridge, Mass.  *General.*  Special, *Orthoptera.  Diurnal Lepidoptera. Fossil Insects.*
1194. Dr. WILLIAM SHARSWOOD, Philadelphia, Pa.  *Coleoptera and Arachnides.*
1195. Prof. HENRY SHIMER, Mt. Carroll, Ill.  *Local.*
1196. GEORGE D. SMITH, No. 162 Washington street, Boston, Mass. *Coleoptera.*
1197. RUFUS SMITH, North Littleton, N. H.  *Local.*
1198. S. I. SMITH (Assistant, Museum of Yale College), New Haven, Ct.  *North American.*  Special, *Orthoptera.*
1199. Dr. WILLIAM M. SMITH, Manlius, N. Y.  *Local.*
1200. CHARLES SONNE, No. 47 La Salle street, Chicago, Ill.  *North American Coleoptera.*
1201. Miss JULIA H. SPEAR, Burlington, Vt.  *Local.*
1202. H. S. SPRAGUE, Buffalo, N. Y.  *Local.*
1203. PHILIP S. SPRAGUE, Dorchester, Mass.  *Local.*
1204. Mrs. PHILIP S. SPRAGUE, Dorchester, Mass.  *Local.*
1205. PHILANDER M. SPRINGER, Springfield, Ill.  *Local.*
1206. SOLOMON STEBBINS, Springfield, Mass.  *General.*  Special, *Diptera.*
1207. E. SUFFERT, Matanzas, Cuba.  *Lepidoptera.*
1208. Prof. F. SUMICHRAST, Orizaba, Mexico.  *Local.*
1209. ROBERT B. TALBOTT, New York, N. Y.  *Local.*
1210. EDWARD TATNALL, jr., Wilmington, Del.  *Coleoptera.*
1211. ALEX. S. TAYLOR, Santa Barbara, Cal.  *Local.*
1212. Prof. SANBORN TENNEY (Professor of Natural Science, Vassar Female College), Poughkeepsie, N. Y.  *Local.*
1213. F. W. TEPPER, Brooklyn, N. Y.  *Local.*
1214. JOHN TEPPER, Brooklyn, N. Y.  *Local.*
1215. D. O. THIEME, Burlington, Iowa.  *Local.*
1216. CYRUS THOMAS, Murphysboro', Ill.  *North American Orthoptera.*
1217. Prof. D. G. THOMPSON (Professor of Natural Sciences, Otterbein University), Westerville, Franklin Co., Ohio.  *Local.*
1218. EDWARD F. TOLMAN (Assistant Curator of Articulata, Worcester Society of Natural History), Worcester, Mass.  *Local.*
1219. Dr. NORTON S. TOWNSHEND, Avon, Lorain Co., Ohio.  *Local.*

1220. JAMES O. TREAT, Lawrence, Mass.  *Local Lepidoptera.*
1221. Dr. ISAAC P. TRIMBLE (State Entomologist of New Jersey), Newark, N. J.  *Local.  Beneficial and Injurious.*
1222. L. TROUVELOT, East Medford, Mass.  *Local.  Special, Silk producing Bombycidæ.  Zoölogical Artist.*
1223. WILLIAM TUPPER, Brooklyn, N. Y.  *Lepidoptera.*
1224. P. R. UHLER (Peabody Institute), Baltimore, Md.  *General.  Special, Hemiptera and Neuroptera.*
1225. HENRY ULKE, Washington, D. C.  *North American Coleoptera.*
1226. Prof. A. E. VERRILL (Professor of Zoölogy, Yale College; Curator of Radiata, Boston Society of Natural History), New Haven, Ct.  *Injurious and Beneficial.*
1227. SAMUEL WAGNER (Editor Bee Gazette), Washington, D.C.  *Apiarian.*
1228. BENJ. D. WALSH, Rock Island, Ill.  *General.  Special, Neuroptera and Cynipidæ.*
1229. Rev. Mr. WASSALL, Newburyport, Mass.  *Lepidoptera.*
1230. J. W. WEIDEMEYER, No. 75 Gold street, New York, N. Y.  *Lepidoptera.*
1231. Mrs. H. W. WELLINGTON, West Roxbury, Mass.  *Local.*
1232. Rev. DAVID WESTON (Curator of Articulata, Worcester Society of Natural History), Worcester, Mass.  *Local.*
1233. C. P. WHITNEY, Milford, N. H.  *Local.*
1234. J. P. WILDE, Egg Harbor, N. J., in summer; Baltimore, Md., in winter.  *Coleoptera.*
1235. —— WILDEBOER, Fontanelle, Barbadoes.  *Local.  A Collector and Dealer.*
1236. Dr. S. C. WILLIAMS, Silver Springs, Lancaster Co., Pa.  *Local.*
1237. CHARLES WILT, No. 1306 South street, Philadelphia, Pa.  *General.*
1238. Prof. ALEXANDER WINCHELL (Professor of Natural History, University of Michigan, Ann Arbor, Mich.  *North American Lepidoptera, Coleoptera, and Hymenoptera.*
1239. Dr. HORATIO C. WOOD, jr. (Professor of Botany, University of Pennsylvania), Academy of Natural Sciences, Philadelphia, Pa.  *North American Myriapoda and Arachnides.*
1240. WILLIAM S. WOOD, No. 61 Walker street, New York, N. Y.  *Coleoptera.*
1241. Rev. DANIEL ZIEGLER, York, Pa.  *North American Coleoptera.*

## CRUSTACEANS.

1242. Prof. LOUIS AGASSIZ (Professor of Zoölogy and Geology, Harvard University; Director and Curator, Museum of Comparative Zoölogy), Cambridge, Mass.  *General.*

1243. CALEB COOKE (Curator of Articulata, Essex Institute), Salem, Mass. *Local.*

1244. Prof. JAMES D. DANA (Professor of Geology and Mineralogy, Yale College), New Haven, Ct. *General.*

1245. C. B. FULLER, Portland, Me. *Local.*

1246. ANDREW GARRETT, care of Samuel Hubbard, San Francisco, Cal. *South Seas.*

1247. Prof. L. R. GIBBES, Charleston, S. C. *Southern Coast.*

1248. Prof. THEODORE GILL (Librarian, Smithsonian Institution), Washington, D. C. *North American.*

1249. C. FRED. HARTT, Cooper Institute, New York, N. Y. *Trilobites.*

1250. Prof. O. C. MARSH (Professor of Palæontology, Yale College), New Haven, Ct. *Fossil.*

1251. F. B. MEEK, Smithsonian Institution, Washington, D. C. *Fossil.*

1252. Gen. ALBERT ORDWAY, Richmond, Va. *General.*

1253. Dr. A. S. PACKARD, jr. (Curator of Crustacea, Boston Society of Natural History; Curator of Articulata, Essex Institute; Editor, American Naturalist), Salem, Mass. *General.* Special, *North Atlantic.* [*Cuban.*

1254. Prof. FELIPE POEY, Calle del Aguila, No. 157, Habana, Cuba.

1255. Dr. EDMUND RAVENEL, Charleston, S. C. *Local and Fossil.*

1256. S. I. SMITH (Assistant in Zoölogy, Yale College), New Haven, Ct. *General.*

1257. Dr. WILLIAM STIMPSON (Secretary, and Director of the Museum, Chicago Academy of Sciences), Chicago, Ill. *General.*

## WORMS.

1258. A. E. R. AGASSIZ (Assistant, Museum of Comparative Zoölogy), Cambridge, Mass. *Marine.*

1259. Prof. JOSEPH LEIDY (Professor of Anatomy, University of Pennsylvania; Curator, Academy of Natural Sciences), No. 1302 Filbert street, Philadelphia, Pa. *Parasites.*

1260. WILLIAM C. MINOR, New Haven, Ct. *Marine, Local.*

1261. Dr. A. S. PACKARD, jr. (Curator of Crustacea, Boston Society of Natural History; Curator of Articulata, Essex Institute; Editor, American Naturalist), Salem, Mass. *North Atlantic.*

1262. Dr. WILLIAM STIMPSON (Secretary and Director of the Museum, Chicago Academy of Sciences), Chicago, Ill. *Marine.*

1263. Dr. F. R. STURGIS, No. 103 Ninth street, New York, N. Y. *Helminths.*

1264. Prof. A. E. VERRILL (Professor of Zoölogy, Yale College; Curator of Radiata, Boston Society of Natural History), New Haven, Ct. *Local.*

1265. Dr. J. C. WHITE (Professor of ———, Mass. Medical College; Curator of Mammalia and Comparative Anatomy, Boston Society of Natural History), Boston, Mass. *Helminths.*
1266. Prof. ALEXANDER WINCHELL (Professor of Natural History, University of Michigan), Ann Arbor, Mich. *Helminths.*

## MOLLUSKS.

1267. Prof. LOUIS AGASSIZ (Professor of Zoölogy and Geology, Harvard University; Curator and Director, Museum of Comparative Zoölogy), Cambridge, Mass. *General.*
1268. TRUMAN H. ALDRICH, Troy, N. Y. *Local.*
1269. ANSON ALLEN, Orono, Me. *Terrestrial and Fluviatile.*
1270. J. G. ANTHONY (Assistant, Museum of Comparative Zoölogy), Cambridge, Mass. *General.* Special, *North American Terrestrial and Fluviatile.*
1271. RAFAEL ARANGO, Habana, Cuba. *Cuban.*
1272. J. W. ARNOLD (Cabinet Keeper and Curator of Mollusca, Worcester Society of Natural History), Worcester, Mass. *Local.*
1273. E. P. AUSTIN, Cambridge, Mass. *Local Terrestrial and Fluviatile.*
1274. Rev. JOSEPH BANVARD, Patterson, N. J. *Local.*
1275. Rev. E. R. BEADLE (Secretary, Conchological Section, Academy of Natural Sciences, Philadelphia, Pa. *General.*
1276. A. S. BICKMORE, New York, N. Y.), *General.*
1277. B. BILLINGS, Ottawa City, Canada West. *Local.*
1278. E. BILLINGS (Palæontologist, Geological Survey of Canada), Montreal, Canada. *Fossil.*
1279. WILLIAM G. BINNEY, Burlington, N. J. *North American Terrestrial.*
1280. THOMAS BLAND, Brooklyn, N. Y., or No. 42 Pine street, New York, N. Y. *North American and West Indian Terrestrial.*
1281. Rev. E. C. BOLLES (Corresponding Secretary, Portland Society of Natural History), Portland, Me. *North American Terrestrial and Fluviatile.*
1282. A. D. BROWN, Princeton, N. J. *Terrestrial.*
1283. ROBERT H. BROWNNE (Recording Secretary, Lyceum of Natural History of New York), No. 54 West Fifteenth street, New York City; No. 91 South Ninth street, Williamsburgh, N. Y. *General.*
1284. Dr. P. P. CARPENTER, Montreal, Canada. *General.* Special, *Pacific Coast of North America.*
1285. RICARDO I. CAY, Matanzas, Cuba. *Cuban.*

1286. L. E. CHITTENDEN, No. 252 Broadway, New York, N. Y. *Local.*
1287. Dr. DANIEL CLARKE (President, Flint Scientific Institute), Flint, Mich. *Local.*
1288. Prof. H. JAMES-CLARK (Professor of Natural History, Pennsylvania Agricultural College), Centre Co., Pa. *Anatomy.*
1289. WILLIAM C. CLEVELAND, No. 46 Washington street, Boston, Mass. *North American Terrestrial and Fluviatile.*
1290. T. A. CONRAD, Academy of Natural Sciences, Philadelphia, Pa. *Naiades and Fossil.*
1291. CALEB COOKE (Curator of Articulata, Essex Institute), Salem, Mass. *Local.*
1292. Dr. FRANCISCO J. CORONADO, Habana, Cuba. *Cuban.*
1293. A. O. CURRIER, Grand Rapids, Mich. *North American.*
1294. WILLIAM H. DALL, Academy of Natural Sciences, Chicago, Ill. *North American.* Special, *Pacific Slope.*
1295. HENRY DAVIS, McGregor, Iowa. *Local.*
1296. Dr. J. W. DAWSON (Principal, McGill University), Montreal, Canada. *Fossil.*
1297. Dr. WM. H. DE CAMP, Grand Rapids, Mich. *North American.*
1298. A. DIETZ, St. Thomas, West Indies. *General.* Special, *West Indian.*
1299. G. W. DUNN, San Francisco, Cal. *General.*
1300. DAVID W. FERGUSON, Brooklyn, N. Y. *Local.*
1301. Dr. A. E. FOOTE, Ann Arbor, Mich. *North American Terrestrial and Fluviatile.*
1302. JONATHAN FORD, Philadelphia, Pa. *Marine Gasteropods.*
1303. Dr. E. FOREMAN, Catonsville, Md. *General.*
1304. R. J. FOWLER, Montreal, Canada. *Local.*
1305. C. B. FULLER, Portland, Me. *Local.*
1306. ANDREW GARRETT, Care of Samuel Hubbard, San Francisco, Cal. *South Seas.*
1307. Prof. THEODORE GILL (Librarian, Smithsonian Institution), Washington, D. C. *General.*
1308. THOMAS A. GREENE, New Bedford, Mass. *General.*
1309. H. HAAGENSEN, St. Thomas, West Indies. *West Indian.*
1310. WILLIAM A. HAINES, No. 177 Madison Avenue, New York, N. Y. *General.*
1311. Prof. S. S. HALDEMAN, Columbia, Pa. *North American Fluviatile.*
1312. Mrs. I. D. HALL, New Bedford, Mass. *Local.*
1313. Prof. JAMES HALL (Curator, State Geological Museum, State Geologist of New York), Albany, N. Y. *Fossil.*
1314. Dr. W. H. HARTMAN, Westchester, Chester Co., Pa. *Local.*
1315. M. W. HARRINGTON (Assistant, Museum of the University of

Michigan), Ann Arbor, Mich. *North American Terrestrial and Fluviatile.*

1316. C. FRED. HARTT, Cooper Institute, New York, N. Y. *Brachiopods, Living and Fossil.*

1317. J. P. HASKELL, Marblehead, Mass. *Local.*

1318. THOMAS C. HASKELL, Swampscott, Mass. *Local.*

1319. Prof. F. V. HAYDEN (Professor of Geology and Mineralogy, University of Pennsylvania), Philadelphia, Pa. *North American Fossil.*

1320. Prof. F. S. HOLMES, Charleston, S. C. *Southern States, Living and Fossil.*

1321. ROBERT HOWELL, Nichols, Tioga Co., N. Y. *Local.*

1322. Dr. S. B. HOWELL (Chairman, Committee on Cephalopoda, Conchological Section, Academy of Natural Sciences), Philadelphia, Pa. *Cephalopods.*

1323. Dr. P. R. HOY, Racine, Wis. *Local.*

1324. Dr. E. W. HUBBARD, Tottenville, Staten Island, N. Y. *General.*

1325. ALPHEUS HYATT (Curator of Palæontology, Boston Society of Natural History; Curator of Polyzoa and Palæontology, Essex Institute; Editor, American Naturalist), Salem, Mass. *Cephalopods and Polyzoa, Living and Fossil.*

1326. J. W. JACKMAN, Newburyport, Mass. *Local.*

1327. U. P. JAMES, Cincinnati, Ohio. *North American.*

1328. Dr. JNO. C. JAY, Rye, Westchester Co., N. Y. *General.*

1329. Col. EZEKIEL JEWETT, Utica, N. Y. *Fossil, and a Collector.*

1330. FRANCISCO DE JIMENO, Matanzas, Cuba. *Cuban.*

1331. Prof. SAMUEL JONES (Professor of Natural Science, Jefferson College), Canonsburg, Washington Co., N. Y. *Local.*

1332. Rev. P. P. KIDDER, Ellcottsville, N. Y. *Local.*

1333. Capt. H. F. KING (Curator of Mollusca, Essex Institute), Salem, Mass. *General Collection.*

1334. Rev. A. B. KENDIG, Davenport, Iowa. *North American Fluviatile and Terrestrial Gasteropods.*

1335. Dr. J. P. KIRTLAND, East Rockport, Ohio; Post-office address, Cleveland, Ohio. *Local.*

1336. HENRY KREBS, St. Thomas, West Indies. *West Indian.*

1337. Dr. I. A. LAPHAM (President, Wisconsin Historical Society), Milwaukie, Wis. *Local.*

1338. Dr. GEORGE A. LATHROP, East Saganaw, Mich. *Local.*

1339. ISAAC LEA (Vice President, American Philosophical Society; Director, Conchological Section, Academy of Natural Sciences), No. 1622 Locust street, Philadelphia, Pa. *Fluviatile, Terrestrial, and Fossil.*

1340. Dr. JAMES LEWIS, Mohawk, N. Y. *Local.*
1341. Rev. SAMUEL LOCKWOOD, Keyport, N. J. *Local.*
1342. A. B. LYON, Ann Arbor, Mich. *Terrestrial and Fluviatile.*
1343. W. L. MACTIER (Treasurer, Conchological Section, Academy of Natural Sciences), No. 132 Walnut street, Philadelphia, Pa. *General.*
1344. WILLIAM T. MARCH, Spanishtown, Jamaica. *West Indian.*
1345. Prof. R. Z. MASON, Appleton, Wis. *Local.*
1346. E. R. MAYO, No. 82 Milk street, Boston, Mass.
1347. F. B. MEEK, Smithsonian Institution, Washington, D. C. *Fossil.*
1348. Dr. MANLY MILES (Professor of Animal Physiology and Practical Agriculture, State Agricultural College), Lansing, Mich. *Local.*
1349. HENRY MOORES, Columbus, Ohio. *Local.*
1350. EDWARD S. MORSE (Curator of Mollusca, Boston Society of Natural History; Curator of Mollusca, Essex Institute; Editor, American Naturalist), Salem, Mass. *General.* Special, *North American Fluviatile and Terrestrial.*
1351. Dr. WESLEY NEWCOMB, San Francisco, Cal. *General.* Special, *Pacific Slope of America. Achatinellæ.*
1352. Dr. EDWARD J. NOLAND (Conservator, Conchological Section, Academy of Natural Sciences), Philadelphia, Pa. *General.*
1353. CHARLES F. PARKER (Librarian, Conchological Section, Academy of Natural Sciences), Philadelphia, Pa. *General.*
1354. GEORGE H. PERKINS, Galesburg, Ill. *North American.*
1355. JNO. S. PHILLIPS, Philadelphia, Pa. *Marine Acephala.*
1356. Prof. J. W. Powell, Normal, Ill. *Fluviatile and Fossil.*
1357. MANUEL J. PRESAS, Calle de Velarde, No. 5, Matanzas, Cuba. *Cuban.*
1358. TEMPLE PRIME, No. 26 Broad street, New York, N. Y. *Corbiculadæ.*
1359. Dr. EDMUND RAVENEL, Charleston, S. C. *Local and Fossil.*
1360. J. R. READ, New Bedford, Mass. *Local.*
1361. J. H. REDFIELD, Philadelphia, Pa. *Local.*
1362. A. B. RICHMOND, Meadsville, Pa. *Local.*
1363. A. H. RIISE, St. Thomas, West Indies. *West Indian.*
1364. T. RIMMER, Montreal, Canada. *General.*
1365. S. R. ROBERTS (Recorder, Conchological Section, Academy of Natural Sciences), Philadelphia, Pa. *General.*
1366. C. T. ROBINSON (Curator of Mollusca, Buffalo Society of Natural Science), No. 31 Wall street, New York, N. Y. *General.*
1367. HENRY ROUSSEAU, Troy, N. Y. *Fossil. Local.*
1368. Rev. J. ROWELL, San Francisco, Cal. *Local.*
1369. Dr. W. S. W. RUSCHENBERGER, Philadelphia, Pa. *General.*

# 57 MOLLUSKS. 1370–1392

1370. Prof. ABRAM SAGER (Professor of Obstetrics, University of Michigan), Ann Arbor, Mich. *Local.*
1371. JAMES SAMPSON, New Harmony, Ind. *Local.*
1372. GEORGE SCARBOROUGH, Sumner, Atchinson Co., Kansas. *Local.*
1373. N. S. SHALER (Assistant, Museum of Comparative Zoölogy), Cambridge, Mass. *Brachiopods, Living and Fossil.*
1374. Prof. D. S. SHELDON (Professor of Chemistry and Natural Sciences, Griswold College), Davenport, Iowa. *Terrestrial and Fluviatile.*
1375. Dr. E. R. SHOWALTER, Uniontown, Ala. *Local.*
1376. SANDERSON SMITH, No. 26 Broad street, New York, N. Y. *Local.*
1377. ROBERT E. C. STEARNS (Vice President and Curator of Mollusca, California Academy of Sciences), Lock box 1449, Post-office, San Francisco, Cal. *General.* Special, *Pacific Slope of America.*
1378. J. B. STEERE (Assistant, Museum of the University of Michigan), Ann Arbor, Mich. *North American Terrestrial and Fluviatile.*
1379. D. JACKSON STEWARD, No. 148 Fifth Avenue, New York, N. Y. *General.*
1380. WM. W. STEWART (Custodian, Buffalo Society of Natural Sciences), Buffalo, N. Y. *Local.*
1381. Dr. WILLIAM STIMPSON (Secretary and Director, Chicago Academy of Sciences), Chicago, Ill. *General.* Special, *Atlantic coast of North America.*
1382. FRANK W. STOWELL, No. 191 Fulton avenue, Brooklyn, N. Y. *American.*
1383. RICHARD H. STRETCH, Virginia City, Nevada. *Local.*
1384. ROBERT L. STUART, No. 154 Fifth avenue, New York, N. Y. *General.*
1385. JOSEPH SULLIVANT, Columbus, Ohio. *Local.*
1386. ROBERT SWIFT, St. Thomas, West Indies, or care of Thomas Bland, 42 Pine street, New York, N. Y. *General.* Special, *West Indian.*
1387. Dr. THEODORE A. TELLKAMPF, No. 142 West Fourth street, New York, N. Y. *Ascidians.*
1388. L. L. THAXTER, No. 13 Tremont street, Boston, Mass. *New England Terrestrial and Fluviatile.*
1389. JOHN H. THOMSON, New Bedford, Mass. *Terrestrial and Fluviatile Gasteropods.*
1390. Dr. J. B. TRASK, San Francisco, Cal. *Californian.*
1391. Dr. J. B. TREMBLEY, Toledo, Ohio. *Local.*
1392. GEORGE W. TRYON, Jr. (Vice Director, Conchological Section, Academy of Natural Sciences; Editor, American Journal of Conchology), No. 625 Market st., Philadelphia, Pa. *General.*

1393. J. C. TURNPENNY, Philadelphia, Pa. *Marine Acephala.*
1394. HENRY D. VAN NOSTRAND, No. 116 West street, New York, N. Y. *General.*
1395. Prof. A. E. VERRILL (Professor of Zoölogy, Yale College; Curator of Radiates, Boston Society of Natural History), New Haven, Ct. *American.*
1396. Rev. Mr. VILLENEUVE, Montreal, Canada. *Local.*
1397. W. E. WELLINGTON, Dubuque, Iowa. *Local.*
1398. CHARLES M. WHEATLEY, Phœnixville, Pa., or 42 Pine street, New York, N. Y. *Fluviatile.*
1399. JOHN M. WHEATON, Columbus, Ohio. *Local.*
1400. J. F. WHITEAVES (Recording Secretary and Curator, Natural History Society of Montreal), Montreal, Canada. *General.*
1401. HENRY S. WILLIAMS, Ithaca, N. Y. *Local, Living and Fossil.*
1402. J. R. WILLIS, Halifax, Nova Scotia. *Local.*
1403. Prof. ALEX. WINCHELL (Professor of Natural History, University of Michigan), Ann Arbor, Mich. *Terrestrial and Fluviatile.*
1404. A. YOUNG, Brooklyn, N. Y. *General.*

## RADIATES.

1405. A. E. R. AGASSIZ (Assistant, Museum of Comparative Zoölogy), Cambridge, Mass. *General.* Special, *Echinoderms and Acalephs.*
1406. Prof. LOUIS AGASSIZ (Professor of Geology and Zoölogy, Harvard University; Curator and Director, Museum of Comparative Zoölogy), Cambridge, Mass. *General.*
1407. Rev. WILLIAM H. BARRIS, Burlington, Iowa. *Fossil Crinoids.*
1408. Prof. H. JAMES-CLARK (Professor of Natural History, Pennsylvania Agricultural College), Centre Co., Pa. *Anatomy.*
1409. Prof. JAMES D. DANA (Professor of Geology and Mineralogy, Yale College), New Haven, Ct. *Polyps.*
1410. P. DUCHASSAING, St. Thomas, West Indies. *Polyps.*
1411. CHARLES B. FULLER, Portland, Me. *New England.*
1412. B. J. HALL, Burlington, Iowa. *Fossil Crinoids.*
1413. Prof. C. FRED. HARTT (Professor of Natural History, Vassar College), Poughkeepsie, N. Y. *Polyps.*
1414. JOHN G. HEYWOOD (Assistant Curator of Radiata, Worcester Society of Natural History), Worcester, Mass.
1415. Dr. J. B. HOLDER, U. S. Army. *Florida Corals.*
1416. Col. THEODORE LYMAN (Assistant, Museum of Comparative Zoölogy), Brookline, Mass. *Ophiurans and Polyps.*

59    RADIATES. PROTOZOA. PARASITES.   1417–1434

1417. Prof. John McCrady, Charleston, S. C.  *Living and Fossil.*
1418. F. B. Meek, Smithsonian Institution, Washington, D. C.  *Fossil.*
1419. W. H. Niles, Cambridge, Mass.  *Crinoids, Living and Fossil.*
1420. Dr. William Stimpson (Secretary, and Director of the Museum, Chicago Academy of Sciences), Chicago, Ill.  *Echinoderms and Polyps.*
1421. Dr. O. Theime, Burlington, Iowa.  *Fossil Crinoids.*
1422. Prof. A. E. Verrill (Professor of Zoölogy, Yale College; Curator of Radiates, Boston Society of Natural History), New Haven, Ct.  *General.* Special, *Polyps and Echinoderms.*
1423. Dr. Charles A. White (State Geologist of Iowa), Iowa City, Iowa.  *Fossil.*

## PROTOZOA.

1424. Prof. H. James-Clark (Professor of Natural History, Pennsylvania Agricultural College), Centre Co., Pa.  *General.*
1425. P. Duchassaing, St. Thomas, West Indies.  *Sponges.*
1426. Alpheus Hyatt (Curator of Palæontology, Boston Society of Natural History; Curator in Natural History Department, Essex Institute; Ed tor, American Naturalist), Salem, Mass.  *General.*
1427. Prof. O. C. Marsh (Professor of Palæontology, Yale College), New Haven, Ct.  *Fossil Sponges.*
1428. Prof. Alexander Winchell (Professor of Natural History, University of Michigan), Ann Arbor, Mich.  *Local.*

## PARASITES.

1429. Prof. H. James-Clark (Professor of Natural History, Pennsylvania Agricultural College), Centre Co., Pa.  *Vegetable.*
1430. Prof. Joseph Leidy (Professor of Anatomy, University of Pennsylvania; Curator, Academy of Natural Sciences of Philadelphia), No. 1302 Filbert street, Philadelphia, Pa.  *General.*
1431. Prof. J. H. Salisbury (Professor of Physiology, Histology, and Cell Pathology, Charity Hospital Medical College), Cleveland, Ohio.  *General.*
1432. Dr. F. R. Sturgis, No. 103 Ninth street, New York, N. Y.  *Human.*
1433. Prof. J. C. White (Adjunct Professor of Chemistry and Lecturer on Diseases of the Skin), Boston, Mass.  *Human.*
1434. Prof. Jeffries Wyman (Professor of Anatomy and Physiology, Harvard University; President, Boston Society of Natural History), Cambridge, Mass.  *General.*

## ADDITIONAL NAMES RECEIVED.

### GEOLOGY.

1435. WILLIAM ANDREWS, Cumberland, Md. *Local.*
1436. Dr. GEO. S. BLAKIE, Nashville, Tenn. *Local.*
1437. W. T. BRIGHAM (Curator of Geology, Boston Society of Natural History), Boston, Mass. *General Collection.* Special, *Volcanic.*
1438. Dr. A. E. FOOTE (Assistant in Mineralogy, University of Michigan), Ann Arbor, Mich. *General Collection.*
1439. Prof. C. FRED. HARTT (Professor of Natural History, Vassar College), Poughkeepsie, N. Y. *North and South American.*
1440. Prof. HENRY Y. HIND (Professor of Physical and Natural Sciences, Trinity College), Toronto, C. W. *British American.*
1441. JAMES HYATT, Bengall, N. Y. *Local.*
1442. Prof. CHARLES A. JOY, Columbia College, New York, N. Y. *North American.*
1443. Prof. JOSEPH LE CONTE (Professor of Chemistry and Mineralogy, University of South Carolina), Columbia, S. C. *North American.*
1444. Dr. GEORGE LITTLE (State Geologist of Mississippi), Oxford, Miss. *North American.*
1445. THOMAS MCFARLANE, Acton Vale, Canada East. *Local.*
1446. Rev. JOHN B. PERRY (Assistant, Museum of Comparative Zoölogy), Cambridge, Mass. *North American.*
1447. GEORGE L. VOSE, Paris Hill, Me. *General.*
1448. HENRY S. WILLIAMS, Ithaca, N. Y. *Local.*

### PHYSICAL GEOGRAPHY.

1449. A. S. BICKMORE, Amherst, Mass. *Asia.*
1450. Prof. JAMES ORTON, Rochester, N. Y. *Andes and Amazons.*

### MINERALS.

1451. EDWARD S. F. ARNOLD, Yonkers, Westchester Co., N. Y.
1452. Dr. GEO. S. BLAKIE, Nashville, Tenn.
1453. Dr. WM. H. DE CAMP, Grand Rapids, Mich.
1454. Prof. SILAS H. DOUGLASS (Professor of Chemistry and Mineralogy, University of Michigan), Ann Arbor, Mich.

1455. Dr. A. E. FOOTE (Assistant in Mineralogy, University of Michigan), Ann Arbor, Mich.
1456. A. P. GARBER, Columbia, Lancaster Co., Pa.
1457. LEVI HAGER, West Hartford, Vt.
1458. Prof. EUGENE W. HILGARD (Professor of Chemistry and Mineralogy, University of Mississippi), Oxford, Miss.
1459. C. C. HITCHCOCK, Ware, Mass.
1460. Prof. EDMUND O. HOVEY (Professor of Chemistry and Geology, Wabash College), Crawfordsville, Ind.
1461. Prof. CHARLES A. JOY, Columbia College, New York, N. Y.
1462. J. B. KEVINSKI, Lancaster, Pa.
1863. Prof. JOSEPH LE CONTE (Professor of Chemistry and Mineralogy, University of South Carolina), Columbia, S. C. *N. American.*
1464. THOMAS MCFARLANE, Acton Vale, Canada East.
1465. Prof. A. E. STRONG, Grand Rapids, Mich.
1466. Prof. G. C. SWALLOW (State Geologist of Missouri and Kansas), Columbia, Boone Co., Mo.
1467. HENRY S. WILLIAMS, Ithaca, N. Y.

## METALLURGY.

1468. Prof. HENRY S. OSBORN (Professor of Mining and Metallurgy, Lafayette College), Easton, Pa.
1469. Dr. ALBERT B. PRESCOTT (Assistant Professor of Chemistry, etc., University of Michigan), Ann Arbor, Mich.

## PALÆONTOLOGY.

1470. WILLIAM ANDREWS, Cumberland, Md. *Local.*
1471. HENRY DAVIS, McGregor, Iowa. *Local. Collector and Dealer.*
1472. W. H. R. LYKINS, Kansas City, Mo. *Local.*
1473. Prof. JOHN MCCRADY, Charleston, S. C. *Southern States.* Special, *Echinoderms and Graptolites.*
1474. EDWARD T. NELSON, New Haven, Ct. *North American.*
1475. G. H. PERKINS, New Haven, Ct. *North American.*
1476. HENRY S. WILLIAMS, Ithaca, N. Y. *Local.*

## ANATOMY AND PHYSIOLOGY.

1477. Prof. CORYDON L. FORD (Professor of Anatomy and Physiology, University of Michigan), Ann Arbor, Mich.
1478. Prof. JOHN LEAMAN (Professor of Human Physiology and Anatomy, Lafayette College), Easton, Pa.

## MICROSCOPY.

1479. Dr. GEO. S. BLAKIE, Nashville, Tenn.
1480. W. H. COBB, Wellsborough, Tioga Co., Pa.
1481. HIRAM A. CUTTING, Lunenburg, Essex Co., Vt.
1482. Dr. J. BAKER EDWARDS, Montreal, Canada.
1483. A. S. RITCHIE, Montreal, Canada.
1484. GEORGE B. SELDEN, Rochester, N. Y.
1485. C. A. SPENCER, Canastota, N. Y.

## BOTANY.

1486. Dr. GEO. S. BLAKIE, Nashville, Tenn. *Local.*
1487. J. J. CARTER, Lyle, Lancaster Co., Pa. *Local.*
1488. A. P. GARBER, Columbia, Lancaster Co., Pa. *Local.*
1489. BENJ. D. GILBERT, Utica, N. Y. *Local.*
1490. E. L. HANKENSON, Newark, N. Y. *Local.*
1491. M. W. HARRINGTON (Assistant in the Museum, University of Michigan), Ann Arbor, Mich. *Local.*
1492. JAMES HYATT, Bengall, N. Y. *Local.*
1493. A. B. LYON (Assistant in the Museum, University of Michigan), Ann Arbor, Mich. *Local.*
1494. D. R. McCORD, Montreal, Canada. *Canadian Ferns.*
1495. L. A. MILLINGTON, Glens Falls, N. Y. *Local.*
1496. Dr. I. S. MOYER, Plumsteadville, Pa. *Local.*
1497. C. S. OSBORNE (Manager, Western Union Telegraph), Suspension Bridge, Niagara Co., N. Y. *Local.*
1498. S. B. PARSONS, Flushing, N. Y. *Local.*
1499. Dr. FRANK SAWERMAN, Apalachicola, Fla. *North American.*
1500. Prof. G. C. SWALLOW (State Geologist of Missouri and Kansas), Columbia, Boone Co., Mo. *Western States.*
1501. Prof. ALEXANDER WINCHELL (Professor of Natural History, University of Michigan), Ann Arbor, Mich. *United States.*

## ARCHÆOLOGY.

1502. SAMUEL R. CARTER, Paris Hill, Oxford Co., Me. *North American.*
1503. HENRY DAVIS, McGregor, Iowa. *Local. Collector and Dealer.*
1504. Dr. WILLIAM H. DE CAMP, Grand Rapids, Mich. *Local.*
1505. C. B. FULLER, Portland, Me. *Local.*
1506. Hon. E. L. HAMLIN (President, Bangor Historical Society), Bangor, Me. *North American.*
1507. Prof. EDWARD HITCHCOCK (Professor of Hygeine and Physical Education, Amherst College). Amherst, Mass. *N. American.*
1508. Rev. N. W. JONES, New York, N. Y. *North American.*
1509. E. S. MORSE (Curator of Mollusks, Boston Society of Natural History; Curator in Natural History Department Essex Institute) Salem, Mass. *New England.*
1510. F. W. PUTNAM (Curator of Fishes, Boston Society of Natural History; Superintendent, Museum of the Essex Institute), Salem, Mass. *New England.*

## ETHNOLOGY.

1511. A. S. BICKMORE, Amherst, Mass. *Asia.*

## MAMMALS.

1512. G. W. ADERHOLD, A. and M. College, Lexington, Ky. *Local. Taxidermist.*

## BIRDS.

1513. G. W. ADERHOLD, A. and M. College, Lexington, Ky. *Local. Taxidermist.*
1514. ANSON ALLEN, Orono, Me. *Local.*
1515. E. A. JOHNSON, Holyoke, Mass. *Local.*
1516. B. F. OWEN, Astoria, Fulton Co., Ill. *Local.*

## FISHES.

1517. Dr. JAMES BLAKE, San Francisco, Cal. *Embiotocoids.*
1518. Dr. J. C. PARKER, Grand Rapids, Mich. *Local.*

## INSEOTS.

1519. EDWARD P. ALLIS, jr., Yellow Springs, Ohio. *Local.*
1520. Prof. ALBERT J. COOK (Professor of Natural History, State Agricultural College), Lansing, Mich. *General Collection.*
1521. E. A. JOHNSON, Holyoke, Mass. *Local.*
1522. T. F. McCURDY, Norwich Town, Ct. *Local.*
1523. Dr. ALEX. F. SAMUELS, Nashotah, Waukesha Co., Wis. *Lepidoptera.*
1524. G. WICKWIRE SMITH (Corresponding Secretary, Kent Scientific Institute), Grand Rapids, Mich. *Local.*
1625. Dr. G. S. WALKER, No. 1226 Washington avenue, St. Louis, Mo. *Local.*

## MOLLUSKS.

1526. Col. F. F. CAVADA, Trinidad, W. I. *West Indian.*
1527. Hon. EDWARD CHITTY, Kingston, Jamaica. *West Indian.*
1528. Dr. J. G. COOPER (Zoölogist, California State Geological Survey), San Francisco, Cal. *Pacific Slope of America.*
1529. W. M. GABB (Palæontologist, California State Geological Survey), San Francisco, Cal. *Fossil.*
1530. W. G. W. HARFORD, San Francisco, Cal. *General.*
1531. Dr. JOHN C. JAY, Mamaroneck, N. Y. *Local.*
1532. GEORGE METZGES, Circleville, Ohio. *Local.*
1533. JOHN A. McNIEL, Grand Rapids, Mich. *General Collector.* Special, *Fresh Water.* (Now collecting in Central America. *Address* care of Peabody Academy of Science, Salem, Mass.)
1534. EDWARD T. NELSON, New Haven, Ct. *Local.*
1535. Dr. R. A. PHILLIPPI, Santiago, Chili. *Chilian.*
1536. Prof. FELIPE POEY, Habana, Cuba. *Cuban.*
1537. HENRY STRENG, Holland, Mich. *Fresh Water. Local.*

## TAXIDERMISTS.

1538. G. W. ADERHOLD, A. & M. College, Lexington, Ky.

1539. JOHN AKHURST, No. 9½ Prospect street, Brooklyn, N. Y.

1540. A. L. BABCOCK, Sherborn, Mass.

1541. J. P. BATES, No. 209½ North Sixth street, St. Louis, Mo.

1542. JOHN G. BELL, No. 339 Broadway, New York, N. Y.

1543. C. L. BLOOD, Corner of Weir and First streets, Taunton, Mass.

1544. Mrs: J. L. BODE, No. 16 North William street, New York, N. Y.

1544a. RUDOLPH BORCHERDT, Chicago, Ill.

1545. C. G. BREWSTER, No. 16 Tremont street, Boston, Mass.

1546. GEORGE E. BROWN, Dedham, Mass.

1547. JOSEPH BRUNO, Philadelphia, Pa.

1548. DE SCHUTE BUCKOW, No. 27 Frankfort street, New York, N. Y.
    *General Collector of South American Animals and Plants.*

1549. WILLIAM COUPER, Henderson's Buildings, Louis street, Quebec,
    Canada.

1550. C. A. CRAIG, Montreal, Canada.

1551. J. C. DEACON, Chicopee, Mass.

1552. T. W. DEWING, Saxonville, Mass.

1553. C. DREXLER, Washington, D. C.

1554. CHARLES FELDMAN, Philadelphia, Pa.

1555. ALEXANDER GALBRAISH, No. 209 North Ninth street, Philadel-
    phia, Pa.

1556. CHARLES GALBRAITH, West Hoboken, N. J.

1557. WILLIAM GALBRAITH, West Hoboken, N. J.

1558. FERDINAND GRUBER, San Francisco, Cal.

1559. GEORGE HENZEL, Lancaster, Pa.

1560. CHARLES A. HOUGHTON, Holliston, Mass.

1561. W. HUNTER, Museum of the Natural History Society, Montreal,
    Canada.

1562. ILGES & SAUTER, No. 15 Frankfort street, New York, N. Y.

1563. JOHN JENKINS, Monroe, Orange Co., N. Y.

1564. SAMUEL JILLSON, Hudson, Mass.

1565. JOHN KRIDER, Corner of Second and Walnut streets, Philadel-
    phia, Pa.

1566. E. V. LORQUIN, San Francisco, Cal.

1567. C. J. MAYNARD, Newtonville, Mass.

1568. GEORGE Y. NICKERSON, No. 42 Williams st., New Bedford, Mass.

1569. G. ORENSHAW, No. 527 North Fifteenth street, Philadelphia, Pa.

1570. JAMES H. ROOME, No. 55 Carmine street, New York, N. Y.

1571. S. H. SYLVESTER, Middleborough, Mass.

1572. JAMES TAYLOR, Philadelphia, Pa.

1573. N. VICKARY, No. 262 Chestnut street, Lynn, Mass.

1574. GEORGE O. WELCH, Washington street, Lynn, Mass.
1575. ALEXANDER WOLLE, Baltimore, Md.
1576. C. J. WOOD, Philadelphia, Pa.

## MINERALOGY.

1577. J. ROSS BROWNÉ, San Francisco, Cal.
1578. W. S. KEYS, San Francisco, Cal.

## COMPARATIVE ANATOMY.

1579. Dr. E. L. LATHROP, Room No. 2 Newberry's Block, Chicago, Ill.

## BOTANY.

1580. JOHN BUTTLE, San Jose, Cal. *Local.*
1581. W. C. CORMACK, New Westminster, British Columbia. *Local.* Special, *Coniferæ.*
1582. LLOYD JONES, Victoria, British Columbia. *Local.* Special, *Ferns.*

## MAMMALS.

1583. Hon. JOHN D. CATON, Ottawa, Ill. *American Cervidæ.*

## BIRDS.

1584. EUGENE V. LORQUIN, San Francisco, Cal. *Local. Taxidermist.*
1585. GEORGE O. WELCH, Washington street, Lynn, Mass. *Local. Taxidermist.*

## INSECTS.

1586. Dr. —— JONES, New Westminster, British Columbia. *Local.* Special, *Lepidoptera.*

## MOLLUSKS.

1587. ANDREW J. BENNETT, Circleville, Ohio. *Local.*

## ARCHÆOLOGY.

1588. Rev. JOSEPH ANDERSON, Waterbury, Ct.
1589. J. J. H. GREGORY, Marblehead, Mass. *North American.*

## MICROSCOPY.

1590. S. A. BRIGGS, Chicago, Ill.

## FISHES.

1591. Dr. J. H. SLACK, Troutdale, Bloomsbury, N. J. *Pisciculture.*

# CORRECTIONS.

## GEOLOGY.

13. FRANK H. BRADLEY. *Change address to* (Assistant, Illinois State Geological Survey), Wilmington, Ill.
23. E. T. COX. Change "*Local*" to *North American.*
30. Rev. E. B. EDDY. *Change address to* Providence, R. I.
32. *Should be* L. Engelbrecht.
35. C. F. ESCHWEILER. *Change address to* Houghton, Mich.
51. Prof. C. H. HITCHCOCK. *Change address to* (Professor of Geology, Lafayette College; State Geologist of Maine and New Hampshire), No. 33 Wall street, New York, N. Y.
63. CLARENCE KING. *Change address to* United States Geologist of the U. S. Geological Expedition of the 40th Parallel.
90. Prof. J. S. NEWBERRY. *Change address to* (Professor of Geology, Columbia College), New York, N. Y.
107. Prof. JAMES M. SAFFORD. *Change address to* Lebanon, Tenn.
117. Prof. S. TENNEY. *Change address to* (Professor of Natural History, Williams College), Williamstown, Mass.
126. Prof. J. D. WHITNEY. *Change address to* (Professor of Geology, School of Mining, Harvard University), Cambridge, Mass.

## MINERALOGY.

143. W. T. BRIGHAM. *Add* (Curator of Geology, Boston Society of Natural History).
166. Rev. E. B. EDDY. *Change address to* Providence, R. I.
171. CHRISTIAN FEBIYER. *Change to* CHRISTIAN FEBIGER.
201. Rev. A. B. KENDIG. *Change address to* Dubuque, Iowa.
244. Prof. S. TENNEY. *Change address to* (Professor of Natural History, Williams College), Williamstown, Mass.

## METALLURGY.

273. Prof. J. D. WHITNEY. *Change address to* (Professor of Geology, School of Mining, Harvard University), Cambridge, Mass.

## PALÆONTOLOGY.

286. FRANK H. BRADLEY. *Change address to* (Assistant, Illinois State Geological Survey), Wilmington, Ill.
305. G. R. GILBERT. *Change to* G. K. GILBERT.
312. Prof. E. W. HILGARD. *Cross out* "(State Geologist of Mississippi)."
319. ALPHEUS HYATT. *Change address to* (Curator of Palæontology, Boston Society of Natural History; Curator in Natural History Department, Essex Institute), Salem, Mass.

## PHYSICAL GEOGRAPHY.

1449. A. S. BICKMORE.  *Change address to* Tenant's Harbor, Me.
399. CLARENCE KING.  *Change address to* United States Geologist of the U. S. Geological Expedition of the 40th Parallel.
419. Prof. J. D. WHITNEY.  *Change address to* (Professor of Geology, School of Mining, Harvard University), Cambridge, Mass.

## COMPARATIVE ANATOMY AND PHYSIOLOGY.

441. Dr. B. G. WILDER.  *Change address to* (Professor of Natural History, Cornell University), Ithaca, N. Y.

## MICROSCOPY.

484. ALPHEUS HYATT.  *Change* "Curator of Mollusca" *to* Curator of Palæontology.
461. *Should be* EDWIN BICKNELL.

## BOTANY.

512. E. P. AUSTIN.  *Change address to* Cambridge, Mass.
517. Dr. JACOB BARRATT.  *Change address to* Middletown, Ct.
540. *Should be* J. BUCHANAN.
550. Prof. P. A. CHADBOURNE.  *Change address to* (President, Wisconsin State University), Madison, Wis.
553. *Should be* Dr. A. CLAPP.
574. *Change to* Dr. ELIAS DIFFENBAUGH, No. 1113 Carlton street, Philadelphia, Pa.
652. JOHN MACOUN.  *Add*, Special, *Carices.*
674. CHARLES F. PARKER.  *Change to* Camden, N. J.  *North American.* Special, *New Jersey.*
696. Dr. J. T. ROTHROCK.  *Change address to* (Professor of Botany, Pennsylvania State Agricultural College), Centre Co., Pa.
729. Prof. S. TENNEY.  *Change address to* (Professor of Natural History, Williams College), Williamstown, Mass.

## ETHNOLOGY.

1511. A. S. BICKMORE.  *Change address to* Tenant's Harbor, Me.
778. Dr. J. C. NOTT.  *Change address to* New York, N. Y.

## BIRDS.

877. D. DARWIN HUGHES.  Change "*Local*" to *North American.*
886. Rev. A. B. KENDIG.  *Change address to* Dubuque, Iowa.
908. CHARLES H. NAUMAN.  *Change address to* No. 195 East King street, Lancaster, Pa.  Add, *Oölogy.*

911. HENRY A. PURDIE. *Change address to* West Newton, Mass., and department to *New England.*

## REPTILES.

966 Dr. B. G. WILDER. *Change address to* Ithaca, N. Y.

## FISHES.

977. Dr. WM. O. AYRES. *Cross out* " (Corresponding Secretary, California Academy of Natural Sciences)."
987. Dr. W. P. GIBBONS. *Change address to* Alameda, Cal.
1002. Dr. WILLIAM A. NASON. *Change address to* Algonquin, McHenry Co., Ill.
1013. Dr. B. G. WILDER. *Change address to* Ithaca, N. Y.

## INSECTS.

1033. JAMES S. BEHRENS. Add, *General Collection.*
1048. —— BOTTIN. *Change to* —— BOTTERI.
1077. THOMAS A. DICKINSON. *Change address to* (Secretary, Androscoggin Natural History Society), Lewiston, Me. *Local. Coleoptera and Lepidoptera.*
1081. WILLIAM H. EDWARDS. *Change address to* Coalburgh, Kanawha Co., West Virginia.
1103. AUGUSTUS R. GROTE. *Change address to* Hastings-upon-Hudson, N. Y., and dep't to *General,* Special, *Lepidoptera.*
1127. WILLIAM KEOSTLIN. *Change to* WILLIAM KŒSTLING.
1151. Dr. WILLIAM A. NASON. *Change address to* Algonquin, McHenry Co., Ill.
1178. C. V. RILEY. *Change address to* (State Entomologist of Missouri), No. 2130 Clark street, St. Louis, Mo. *North American. Special, Injurious to Vegetation.*
1203. PHILIP S. SPRAGUE. *Change address to* No. 141 Broadway, South Boston, Mass., and dep't to *North American Coleoptera.*
1204. *Cross out* Mrs. PHILIP S. SPRAGUE.

## MOLLUSKS.

1276. A. S. BICKMORE. *Change address to* Tenant's Harbor, Me.
1289. WILLIAM C. CLEVELAND. *Change address to* (Professor of Engineering, Cornell University), Ithaca, N. Y.
1334. Rev. A. B. KENDIG. *Change address to* Dubuque, Iowa.
1532. GEORGE METZGES. *Should be* GEORGE METZGER.
1351. Dr. WESLEY NEWCOMB. *Change address to* Oakland, Cal.

## DECEASED.

97. (Geology.)  Prof. E. J. PICKETT, Rochester, N. Y.  *Died* October, 1866.

333. (Palæontology.)  998. (Fishes.)  Dr. R. P. MANN, Milford, Ohio.  *Died* ——.

354. (Palæontology.)  AUGUSTE RÉMOND, San Francisco, Cal.  *Died* 1867.

376. (Physical Geography.)  Prof. A. D. BACHE, Washington, D. C.  *Died* 1866.

483. (Microscopy.)  WILLIAM W. HUSE, Brooklyn, N. Y.  *Died* 1867.

572. (Botany.)  W. W. DENSLOW, New York, N. Y.  *Died* 1868.

573. (Botany.)  Prof. CHESTER DEWEY, Rochester, N. Y.  *Died* December, 1867.

604. (Botany.)  1308. (Mollusks.)  THOMAS A. GREENE, New Bedford, Mass.  *Died* 1868.

# INDEX

# NATURALISTS' DIRECTORY.

## NORTH AMERICA AND THE WEST INDIES.

Aaron, Charles E., 1016.
Abbott, C. C., 970.
Abbott, Henry L., 375.
Adee, Alvey A., 1017.
Aderhold, G. W., 1512, 1513, 1588.
Agassiz, A. E. R., 276, 452, 971, 1018, 1258, 1405.
Agassiz, Louis, 1, 275, 453, 784, 953, 972, 1019, 1242, 1267, 1406.
Aiken, Wm. E. A., 2, 504.
Akhurst, John, 814, 1020, 1539.
Alcott, W. P., 505, 1021.
Aldrich, Truman H., 1268.
Aligny, Henry d'., 3, 181.
Allan, George S., 457.
Allen, Anson, 1269, 1514.
Allen, Harrison, 423, 790.
Allen, J. A., 815, 1022.
Allen, Lizzie B., 506.
Allen, Oscar D., 132, 259.
Allen, T. F., 507.
Allis, Edward P., jr., 1519.
Alma, G., 816, 973.
Ambrose, John, 817, 974.
Anderson, C. L., 508.
Anderson, Joseph, 1588.
Andrews, E. B., 4.
Andrews, T. L., 509.
Andrews, Wm., 1435, 1470.
Angus, James, 1023.
Anthony, J. G., 1270.
Antisell, T. L., 510.
Arango, Rafael, 1271.
Arnold, Edward S. F., 1451.
Arnold, J. W., 1272.
Arnold, J. W. S., 458.
Ashburner, William, 260.

Ashton, T. B., 1024.
Atkins, Charles G., 975.
Atwood, N. E., 976.
Austin, C. F., 511.
Austin, E. P., 512 (see corrections, p. 68), 1025, 1273.
Auxer, Samuel, 1026.
Ayres, W. O., 977 (see corrections, p. 69).

### B.

Babcock, Amory L., 818, 1027, 1540.
Bache, A. D., 376. (Deceased. See p. 70.)
Bachman, John, 773, 790½.
Bacon, Austin, 5, 513, 1028.
Baer, O. P., 277.
Bagg, M. M., 514.
Bagwell, G. H., 377.
Bailey, L. W., 6, 459, 515.
Bailey, S. C. H., 133.
Baird, S. F., 785, 791, 819, 954, 978.
Balch, D. M., 7.
Bannister, Henry M., 278.
Banvard, Joseph, 1274.
Barden, Edward E., 134.
Barden, James E., 279.
Barnard, John G., 378.
Barnard, Vincent, 135, 516, 820, 1029.
Barnes, Wm., 136.
Barnston, George, 792, 821.
Barratt, Jacob, 517 (see corrections, p. 68).
Barris, Joseph S., 280.
Barris, Wm. H., 281, 1407.
Bassett, Homer F., 1030.

# NORTH AMERICA AND THE WEST INDIES. 73

Bumstead, F. J., 543.
Bunker, Robert, 1059.
Burgess, Edward, 1060.
Burk, Isaac, 544.
Burton, A. R., 148.
Bushee, James, 149.
Buttle, John, 1580.

## C.

Cabot, J. Elliott, 835.
Cabot, Samuel, 836.
Calverly, Stephen, 1061.
Campbell, R. A., 837.
Canby, Wm. M., 545.
Canfield, C. A., 546.
Canfield, G. A., 793.
Cardeza, John, 150.
Carey, William W., 1062.
Carleton, James H., 382.
Carley, S. T., 289.
Carpenter, P. P., 1284.
Carr, Ezra L., 15.
Carson, Joseph, 547.
Carter, Francis, 548.
Carter, J. J., 1487.
Carter, Samuel R., 16, 151, 1502.
Case, L. B., 290.
Cashman, David A., 1063.
Cassells, J. Lang, 549.
Cassin, John, 838.
Caton, John D., 1583.
Cavada, F. F., 1526.
Cay, Ricardo I., 1285.
Chadbourne, P. A., 550 (see corrections, p. 68).
Chapman, A. W., 551.
Chapman, Edward J., 18, 153, 383.
Chapman, H. H., 1064.
Chandler, Charles F., 17, 152.
Chase, Charles A., 291.
Chase, Joseph E., 1065.
Cheney, T. Apoleon, 19, 292, 756.
Chickering, J. W., 552.
Childs, Chandler, 20.
Chittenden, L. E., 1286.
Chitty, Edward, 1527.
Choate, Isaac B., 154.
Christ, Richard, 839.
Christy, David, 293.
Churchill, Joseph R., 1599.
Chute, A. P., 155.          [p. 68].
Clapp, A., 553 (see corrections,
Clapp, F. A., 1066.
Clark, Daniel, 554.

Clark, H. James, 424, 447, 454, 465, 555, 1288, 1408, 1424, 1429.
Clark, James H., 556.
Clark, Mary H., 557.
Clark, W. S., 156, 558.
Clarke, Daniel, 1287.
Clarke, Samuel C., 840, 982 (see corrections, p. 84).
Clay, Joseph A., 157.
Clementi, V., 1067.
Cleveland, Wm. C., 1289 (see corrections, p. 69).
Clinton, George W., 559.
Coates, M. II., 158.
Cobb, C., 294.
Cobb, W. II., 1480.
Coffin, F. G., 159.
Colton, John, 841.
Colvin, Richard, 1068.
Commons, A., 560.
Conrad, T. A., 295, 1290.
Cook, Albert J., 1520.
Cook, George H., 21.
Cooke, Caleb, 467, 1069, 1243, 1291.
Cooke, J. P., 160.
Cooley, ——, 561.
Cooper, J. D., 983.
Cooper, J. G., 562, 794, 842, 1528.
Cope, Edward D., 795, 955, 984.
Copeman, Arthur S., 466.
Cormack, W. C., 1581.
Coronado, Francisco J., 1292.
Cottle, Thomas, 843.
Coues, Elliott, 796, 844.
Couper, William, 845, 1070, 1549 (see corrections, p. 84).
Court, I., 563.
Cowles, Sylvester, 22, 296.
Cox, E. T., 23 (see corrections, p. 67).
Craig, C. A., 846, 1550.
Crawford, S. W., 384.
Cresson, E. T., 1071.
Croft, Henry, 1072.
Crosier, E. S., 24, 564.
Cross, Osborne, 385.
Currier, A. O., 1293.
Currier, Charles, 1073.
Currier, W. O., 1074.
Curtis, M. A., 565.
Curtiss, A. II., 566.
Cutting, Hiram A., 25, 1481.

## D.

Dahl, Christian, 567.

While the Index has been going through the press, the following additions and corrections have been received.

# ADDITIONS.
## ETHNOLOGY.
1592. Rev. Joseph Anderson, Waterbury, Ct. *American Indians.*

## MOLLUSKS.
1593. Harrison E. Webster, Union College, Schenectady, N. Y.
1594. George Y. Nickerson, 36 Williams street, New Bedford, Mass. *Dealer.*
1595. Henry Freedley, Norristown, Pa. *General Collection.*

## MINERALS.
1596. R. H. Stretch (State Mineralogist of Nevada), Virginia City, Nevada.

## INSECTS.
1597. Henry Edwards, Metropolitan Theatre, San Francisco, Cal. *General Collection.*
1598. W. V. Andrews, West Hoboken, N. J. *Lepidoptera.*
1599. Joseph R. Churchill, Milton Lower Mills, Mass. *New England.*

## MAMMALS.
1600. Dr. Francis R. Stæhli, (Assistant, Museum of Comparative Zoölogy), Cambridge, Mass. *General.*

## BIRDS.
1601. Dr. Francis R. Stæhli (Assistant, Museum of Comparative Zoölogy), Cambridge, Mass. *General.*
1602. Miss Grace Anna Lewis, Sunnyside, Kimberton, Chester Co., Penn.

## TAXIDERMISTS AND DEALERS.
1603. Wallace and Hollingsworth, No. 14 North William street, New York, N. Y.

# CORRECTIONS.

13, 286. FRANK H. BRADLEY. *Change to* (Professor of Natural Science, Hanover College), Hanover, Ind.

840, 982. SAMUEL C. CLARKE, *Change address to* Jamaica Plain, Mass.

845, 1070, 1549. WILLIAM COUPER. *Change to* Ottawa City, O., Canada.

471, 581. ARTHUR M. EDWARDS. *Change to* (Professor of Inorganic and Organic Chemistry, Woman's Medical College), No. 126 Second avenue, house 49 Jane street, New York, N. Y.

861, 1555. Should be ALEXANDER GALBRAITH.

310, 1249, 1316, 1413, 1439. C. FRED HARTT. *Change to* (Professor of Geology, Cornell University), Ithaca, N. Y.

70, 330, 644. Prof. LEO LESQUEREUX. *Change address* to Museum of Comparative Zoölogy, Cambridge, Mass.

650. Should be Dr. STARLING LORING.

1144. THEODORE L. MEAD. Change department to *North American and European Lepidoptera.*

92, 342. W. H. NILES. *Change address to* Cambridge, Mass.

734. Should be NORTON S. TOWNSHEND.

1229. Rev. J. WASSALL. *Change address to* Mazo Manie, Wis.

967, 1401, 1448, 1467, 1476. HENRY S. WILLIAMS. *Change to* (Assistant in Geology, Yale College), New Haven, Ct.

# DECEASED.

625. (Botany.) 1120. (Insects.) Rev. JAMES HUBBERT, R'chmond, Canada East. *Died* 1868.

*The following additions and corrections have been received since the index was issued.*

# ADDITIONS.

1604. Miss RACHEL L. BRODLEY, No. 1015 Cherry street, Philadelphia, Pa. *Botany.*
1605. W. R. LIMPERT, Groveport, Franklin Co., Ohio. *Birds. Taxidermist.*
1606. GEORGE W. LINCECUM, Long Point, Washington Co., Texas. *Insects. Birds.* Local.
1607. E. WILKINSON, jr., Mansfield, Ohio. *Birds. Taxidermist.*

# CORRECTIONS.

696. Dr. J. T. ROTHROCK. McVeytown, Pa., is now the correct address.
69. Prof. J. P. LESLEY'S address is 288 South 3d street, Philadelphia.
1497. C. S. OSBORNE'S address is now Box 12, Rochester, N. Y.
1294. W. H. DALL'S address is now Smithsonian Institution, Washington, D C.
1276, 1449, 1511. Prof. A. S. BICKMORE'S address is now Madison College, New York.
758. Dr. SAMUEL A. GREEN. Leave off the final "e."

# DECEASED.

653. (Botany.) HORACE MANN, Cambridge, Mass. *Died* Nov. 11, 1868.*
838. (Birds.) JOHN CASSIN, Philadelphia, Pa. *Died* Jan. 10, 1869.*

* For obituary notices, see "Bulletin of Essex Institute," Vol. I, Nos. 1 and 2, Jan. and Feb., 1869.

# APPENDIX

## TO THE

## NATURALISTS' DIRECTORY.

It is proposed to issue several pages of this Appendix with each number of the Proceedings, in order to allow naturalists an early opportunity of stating their wants and what specimens they have for sale or exchange; to give notices of proposed works on Natural History; changes of address of persons whose names have appeared in the Directory; short obituary notices; new names for the Directory; and such other matters as may be of interest and appropriate to the work.

Notices of the change of address or the decease of any person whose name has appeared in the Directory are especially requested, and any person knowing Naturalists whose names have been omitted, will confer a favor by sending the addresses of such to the editor.

Five lines in the "Appendix" are allowed to each subscriber to the Proceedings for notices of specimens and articles for sale. When a notice of more than five lines is inserted 10 *cents per line* will be charged for every additional line. Non subscribers will be charged 10 *cents per line* for any advertisement of specimens for sale.

All notices from subscribers or others, except of specimens or articles for sale, inserted free, at the discretion of the editor.

F. W. PUTNAM,

Editor.

ESSEX INSTITUTE, June 18, 1866.

# DESIDERATA

### AND

## SPECIMENS FOR SALE AND EXCHANGE.

---

### GEOLOGY AND MINERALOGY.

PROF. ROBERT BELL, Queen's University, Kingston, C. W.

Canadian Minerals and Geological Specimens for exchange.

HENRY D'ALIGNY, Houghton, L. S., Mich.

Has for exchange extensive collections of specimens illustrating the Geology and Mineralogy of the state of Michigan, especially of the Lake Superior Region. Gypsum, Saginaw Bay Salt, Silver Lead, Specular Iron, Pyrolusite, Hematite, Magnetic Iron, Native Silver, Native Copper, Characteristic Vein Stone, Cupriferous Amygdaloids, Cupriferous Conglomerates, &c. Rare specimens, *Whitneyite and Chlorastrolite*.

PROF. HENRY HOW, King's College, Windsor, N. S.

Minerals for exchange.

ISAAC C. MARTINDALE, Byberry, Pa.

Minerals for exchange.

REV. A. B. KENDIG, Marshalltown, Marshall Co., Iowa.

Minerals for exchange.

REV. E. N. BARTLETT, Oberlin, Ohio.

Has a few Geodes from Hamilton, Ill., for sale.

J. D. PARKER, Steuben, Me.

Minerals for exchange.

CHR. C. BROOKS, 53 St. Paul street, Baltimore, Md.

Minerals for exchange.

W. M. HUNTING, Fairfield, Herkimer Co., N. Y.

Quartz Crystals from Middleville, N. Y. for sale, or exchanged for other minerals.

JOHN JENKINS, Monroe, Orange Co., N. Y.

Minerals for sale.

W. W. JEFFERIS, Westchester, Pa.

Minerals for exchange.

SAMUEL R. CARTER, Paris Hill, Oxford Co., Me.

Cabinet specimens of the Minerals occuring at Mount Mica, Paris (Me.), Mount Rubéllite, Hebron (Me.), and vicinity, exchanged for Minerals and Fossils from other localities.

WINSLOW J. HOWARD, 345 Grand street, New York, N. Y.

Rocky Mountain Minerals for exchange.

VINCENT BARNARD, Kennett Square, Chester Co., Pa.

Geological specimens for exchange.

HIRAM A. CUTTING, Lunenburg, Vt.

Geological specimens for exchange.

Dr. Theo. A. Tellkampf, 142 West 4th street, New York, N. Y.
Wishes to obtain a few good cabinet specimens of Beryl.
E. Seymour, 52 Beekman street, New York, N. Y.
Dealer in Minerals.
M. Fox & Co. 10 Courtlandt street, New York, N. Y.
Dealer in Minerals.
C. W. A. Herrmann, 607 Broadway, New York, N. Y.
Dealer in Minerals.
James Eights, Albany, N. Y.
Dealer in Minerals.
C. G. Brewster, 16 Tremont street, Boston, Mass.
Dealer in Minerals.

## FOSSILS.

Prof. F. S. Holmes, College of Charleston, Charleston, S. C.
South Carolina Tertiary Fossils for exchange.
John Gebhard, Jr., Schoharie, N. Y.
Offers his Palæontological collection for sale.
Prof. Henry How, King's College, Windsor, N. S.
Fossils for exchange.
Prof. Robert Bell, Queen's University, Kingston, C. W.
Canadian Fossils for exchange.
J. G. Batterson, Hartford, Ct.
Dealer in Fossils.
C. G. Brewster, 16 Tremont street, Boston, Mass.
Dealer in Fossils.
John Jenkins, Monroe, Orange Co., N. Y.
Fossils for sale.

## BOTANICAL SPECIMENS.

Elihu Hall, Athens, Menard Co., Ill.
Rocky Mountain and Western Plants for exchange for other American species.
Dr. J. W. Robbins, Uxbridge, Mass.
Plants collected at the White Mountains, coast of Connecticut, Vermont, Maryland, Virginia, Lake Superior, Texas and Cuba, for exchange.
Dr. Daniel Clark, Flint, Mich.
Botanical specimens from Michigan for exchange.
Wants a collection of the Alpine Plants of the White Mountains.
Vincent Barnard, Kennet Square, Chester Co., Pa.
Botanical specimens for exchange.
Isaac C. Martindale, Byberry, Pa.
Botanical specimens for exchange.
Thure Kumlien, Busseyville P. O. via Albion, Wis.
Botanical specimens for sale or exchange.

Prof. Robert Bell, Queen's University, Kingston, C. W.
Canadian Plants for exchange.
John Macoun, Bellville, C. W.
Canadian Plants for exchange.
Winslow J. Howard, 345 Grand street, New York, N. Y.
A large collection of Botanical specimens, collected during three years research at the Rocky Mountains, for sale.
Rocky Mountain Plants for exchange.
Dr. F. J. Bumstead, 162 West 23d street, New York, N. Y.
New Jersey, New York and White Mountain Plants for exchange for Plants from other parts of the country.
Rev. Joseph Blake, Gilmanton, N. H.
Botanical specimens for exchange.

## ZOÖLOGICAL SPECIMENS IN GENERAL.
Dr. Daniel Clark, Flint, Mich.
Local Zoölogical specimens for exchange.
Vincent Barnard, Kennet Square, Chester Co., Pa.
Local Zoölogical specimens for exchange.
James Lewis, Mohawk, N. Y.
Has a large general collection for sale.
Prof. Robert Bell, Queen's University, Kingston, C. W.
Canadian Zoölogical specimens for exchange.

## MAMMALS.
See list of Taxidermists, most of whom have skins of mammals for sale or exchange, or can procure specimens.

## BIRDS AND EGGS.
See list of Taxidermists, most of whom have skins and eggs on hand.
D. G. Guimares, New York, N. Y.
Dealer in South American Skins.
John Krider, Corner 2d and Walnut streets, Philadelphia, Pa.
Birds' skins and Eggs for sale and exchange.
A collection of North American Birds, consisting of 483 species and about 1000 specimens, for sale. Price $1200. The skins are in the best condition. Also a collection of Birds' Eggs, consisting of 290 species, 750 specimens. Price $500. Catalogues furnished on application.
S. Jillson, Feltonville, Mass.
Native Birds' skins for sale or exchange.
A. L. Babcock, Sherborn, Mass.
Native Birds' skins for exchange.
Wants species from the Southern States, South America and the West Indies; especially Humming Birds, Parrots and Chatterers.
L. J. Maynard, Newtonville, Mass.
Collects local specimens for sale.

Dr. Wm. Wood, East Windsor Hill, Ct.
Skins and Eggs of North American Birds for exchange.
A collection of 600 mounted Birds for sale.

Thure Kumlien, Bussyville P. O. via Albion, Wis.
Skins and Eggs of Birds for sale or exchange.

Rev. A. B. Kendig, Marshalltown, Marshall Co., Iowa.
Birds for exchange.

B. P. Mann, Cambridge, Mass.
Birds' Eggs, principally North American, for sale or exchange; exchange preferred.

E. A. Samuels, State Collection, Boston, Mass.
Eggs of New England Birds for exchange.

Wm. Couper, Quebec, Canada.
Eggs of the following Birds for sale: *Archibuteo lagopus*, *Surnia ulula*, *Brachyotus Cassinii*, *Saxicola œnanthe*, *Ægiothus linarius*, *Corvus Americanus*, *Pica Hudsonica*, *Lagopus Americanus*, *Spatula clypeata*, *Chaulelasmus streperus*, *Histrionicus torquatus*, *Melanetta velvetina*.

D. Andrade & Co., 38 Walker street, New York, N. Y.
Himalayan Pheasant skins for sale.

INSECTS.

James Ridings, 1311 South street, Philadelphia, Pa.
Named specimens of North American Coleoptera for sale.

E. Suffert, Matanzas, Cuba.
Cuban Lepidoptera for exchange for Lepidoptera from other parts of the world.

Dr. Herrick-Schaeffer, Regensburg, Bavaria.
Has large collections of European Lepidoptera, especially Microlepidoptera, and Geometrids, well prepared and correctly determined, which he can furnish in any quantity up to 1000 species in a lot, either by sale or in exchange for American specimens. He especially wishes the Californian and American Heterocera.

Deyrolle et Fils, Naturalistes, 19 Rue de la Monnaie, Paris, France.
Large collections of named Coleoptera from Brazil, Antilles, Venezuela, New Grenada, and Guyana, for sale. Priced catalogues furnished on application.

Wm. Couper, Quebec, Canada.
Wants to procure the following Coleoptera : *Omus Californicus* Esch., *Amblycheila cylindriformis* Say, *Cicindela splendida* Hentz, *C. trifasciata* Fabr., *C. obsoleta* Say, *C. pusilla* Say, *C. terricola* Say, *Lachnophorus elegantulus* Mann., *Lebia grandis* Hentz, *L. pulchella* Dej., *Coptodera signata* Dej., *Thalpius pygmœus* Dej., *Drepanus LeContei* Dej., *Scarites quadriceps* Chaud., *Pasimachus punctulatus* Hald., *Cychrus velutinus* Mén., *C. tuberculatus* Harris, *Scaphinotus heros* Harris, *S. elevatus* Fabr., *Sphœroderus bilobus* Say, *Notiophilus semistriatus* Say, *Badister notatus* Hald., *Dytiscus confluens* Say, *D. Harrisii* Kirby, *Ne-*

*crophorus Americanus* Oliv., *N. mediatus* Fabr., *N. marginatus* Fabr., *N. pustulatus* Illig., *Lucanus elaphus* Fabr., *Buprestis sexnotata* Lap., *B. lineata* Fabr., *B. fasciata* Fabr., *B. decora* Fabr., *B. characteristica* Harris. Offers to exchange Northern Coleopterous Insects or North American Birds' Eggs for the above.

WILSON ARMISTEAD, Virginia House, Leeds, England.

This gentleman is engaged upon a work on the Galls and Gall Insects, and requests specimens and observations from all parts of the world.

GEORGE C. BRACKETT, Belfast, Me.

Local Insects for sale.

JOHN AKHURST, 9½ Prospect street, Brooklyn, N. Y.

Coleoptera and Lepidoptera for exchange.

ISAAC A. POOL, 829 Washington street, Chicago, Ill.

Local Insects for exchange.

W. W. CARY, Coleraine, Mass.

Italian Bees for sale.

DR. A. S. PACKARD, Boston Society of Natural History.

Solicits nests containing the larvæ and pupæ of Mud Wasps, Sand Wasps, Paper Wasps; colonies of Humble Bees, Wild Bees, &c. Any information relative to the habits, the mode of making their cells and nests, or to the parasites of the above insects, will be duly accredited. Especial attention is called to the nests or cells of the Solitary Bees which make deep holes in sunny paths and banks; to the Wasps which tunnel the stems of the Dock, Elder, Currant, Raspberry, Blackberry, Syringa, &c.; to those species of Odynerus, etc., which construct mud cells in deserted galls and nests of other insects. If desired, specimens received will be returned carefully labelled. The larvæ and pupæ should be preserved in whiskey (high wines) or glycerine, and the perfect insects should be pinned. All the specimens should be carefully labelled with the exact locality, date of collection, and reference to the nest from which they were taken. Dr. Packard's paper on the larvæ and pupæ of the Hymenoptera is to be printed, with figures, in the Proceedings of the Essex Institute.

## MOLLUSCA.

WM. HARPER PEASE, Honolulu, Sandwich Islands.

Wants Marine Gasteropods, especially from the East Coast of Africa, Red Sea, East Indies and West Indies. Offers in exchange Land, Freshwater and Marine Shells of the Pacific Islands.

R. E. C. STERNS, Lock box 1449, San Francisco, Cal.

Pacific coast and Californian Shells for exchange for species from other localities.

DR. GEORGE A. LATHROP, East Saginaw, Mich.

Has Land and Fresh-water Shells of his vicinity for exchange for Shells from other localities, Marine species included.

ANSON ALLEN, Orono, Me.

Local species of Land and Fresh-water Shells for exchange.

GEORGE SCARBOROUGH, Sumner, Atchinson Co., Kansas.

Ohio River Shells for exchange.

REV. A. B. KENDIG, Marshalltown, Marshall Co., Iowa.

Land and Fresh-water Gasteropods for exchange.

CHARLES M. WHEATLEY, Phœnixville, Pa.

Fluviatile Mollusks for exchange.

Wants, either by exchange or purchase, species of Fluviatile Mollusks not in his collection.

REV. JOSEPH BANVARD, Worcester, Mass.

Shells for sale.

ISAAC A. POOL, 829 Washington street, Chicago, Ill.

Western Fresh-water Shells for exchange.

ELIHU HALL, Athens, Menard Co., Ill.

North American Land and Fresh-water Shells for exchange for marine species.

REV. E. C. BOLLES, Portland, Me.

New England Shells for exchange for Land and Fresh-water species from other localities.

JOHN GREGORY, 116 South street, New York, N. Y.

Shells for sale.

## GENERAL DEALERS IN SPECIMENS OF NATURAL HISTORY.

C. G. BREWSTER, 16 Tremont street, Boston, Mass.

GEO. Y. NICKERSON, 42 William street, New Bedford, Mass.

JOHN G. BELL, 339 Broadway, New York, N. Y.

————Kaempfer, Madison street, Chicago, Ill.

JAMES EIGHTS, Albany, N. Y.

JOHN AKHURST, 9½ Prospect street, Brooklyn, N. Y.

ILGES AND SANTER, 15 Frankfort street, New York, N. Y.

D. BOURGET, 115 Rua d' Ouvidor, Rio Janeiro, Brazil.

BUFFON & WILSON, 391 Strand, London, W. C., England.

BRYCE M. WRIGHT, 36 Great Russell street, Bloomsbury, London, W. C., England.

ROBERT DAMON, Weymouth, Dorsetshire, England.

B. JACOBS, 68 Leadenhall street, London, England.

JAMES CARFRAE, JR., 79 Princess street, Edinburgh, Scotland.

————BOEHNER, Berlin, Prussia.

A. KOENEN, Berlin, Prussia.

L. W. SCHAUFUSS, Dresden, Saxony.

DEYROLLE ET FILS, 16 Rue de la Monnaie, Paris, France.

EDWARD VERREAUX, Paris, France.

L. PARZUDAKI, Paris, France.

Madame H. DROUET, Paris, France.

ARTHUR ELOFFER, 20 Rue de l'Ecole de Medecine, Paris, France.

JOHN S. STEVENS, Natural History Agency Office, 24 Bloomsbury street, London, W. C., England.

Mr. Stevens is agent for the sale of collections made in various parts of the world, especially from the following places : Europe, Cape of Good Hope, Natal, Gold Coast, Damara Land, Zambesi, Old Calabar, the Gaboon, Madagascar, Bogota, Upper Amazons, Santa Martha, Bahia, Nicaragua, Mexico, Texas, New Guinea, Borneo, Celebes, Sumatra, Java, and other East Indian Islands, Siam, Penang, Cambodia, Laos, Birmah, Himalayan mountains and other parts of India, Ceylon, China, Japan, Philippine Islands, Swan River, South Australia, Victoria, New South Wales, Queensland, North Australia, New Hebrides, Fegee Islands, and New Zealand. Mr. Stevens will send specimens for selection.

## DEALERS IN NATURALISTS' APPARATUS.

AMERICAN NET AND TWINE MANUFACTURING CO., 43 Commercial street, Boston, Mass.

Nets and Seines of all description made to order.

CODMAN AND SHURTLEFF, 13 Tremont street, Boston, Mass.

Manufacturers of Forceps, Knives and other Instruments used by Naturalists.

NEW ENGLAND GLASS CO., Boston, Mass.

Glass Jars of all sizes, with glass stoppers, on hand or made to order, for the preservation of alcoholic specimens.

DONNELL & MOORE, Old Cambridge, Mass.

Manufacturers of Tin and Copper Cans for alcoholic specimens.

J. L. BODE, 16 North William street, New York, N. Y.

Manufacturer of Birds' eyes.

C. G. REWSTER, 16 Tremont street, Boston, Mass.

Insect Pins.

THEODORE SCHRECKEL, North William street, New York, N. Y.

Has on hand and imports Insect Pins and Entomological Apparatus.

F. W. CHRISTERN, Broadway, New York, N. Y.

Imports Insect Pins.

CHARLES STODDER, 75 Kilby street, Boston, Mass.

Agent for R. B. Tolles, J. Zentmeyer, and W. Wales; Opticians and Makers of Microscopes.

H. M. RAYNOR, 748 Broadway, New York, N. Y.

Platinum Apparatus; Tube, Sheet, Wire, &c., in all forms, for all purposes. Wholesale and Retail.

# MICROSCOPE MAKERS.

W. Wales, Fort Lee, Bergen Co., N. J.
G. Wale, Bull's Ferry, Bergen Co., N. J.
J. Grunow, New York, N. Y.
J. Zentmeyer, Philadelphia, Pa.
R. B. Tolles, Canastota, Madison Co., N. Y.
T. II. McAllister, New York, N. Y.

# TAXIDERMISTS.

C. G. Brewster, 16 Tremont street, Boston, Mass.
N. Vickary, 262 Chestnut street, Lynn, Mass.
A. L. Babcock, Sherborn, Mass.
Samuel Jillson, Hudson (Feltonville), Mass.
S. H. Sylvester, Middleborough, Mass.
L. J. Maynard, Newtonville, Mass.
J. C. Deacon, Chicopee, Mass.
C. L. Blood, Corner of Weir and First streets, Taunton, Mass.
George Y. Nickerson, 42 Williams street, New Bedford, Mass.
John Jenkins, Monroe, Orange Co., N. Y.
John G. Bell, 339 Broadway, New York, N. Y.
J. L. Bode, 16 North William street, New York, N. Y.
John Akhurst, 9½ Prospect street, Brooklyn, N. Y.
John Krider, Corner 2d and Walnut streets, Philadelphia, Pa.
George Hensel, Lancaster, Pa.
C. Drexler, Washington D. C.
Alex. Wolle, Baltimore, Md.
Wm. Couper, Henderson's Buildings, Louis street, Quebec, Canada.

# LABELS FOR CABINET SPECIMENS.

The Smithsonian Institution has printed labels of the Family names of American Birds and Mammals, giving both the Scientific and English names. Also special labels for many Birds, Mammals and other specimens, which it will supply at cost. It will also furnish, at cost, its Check Lists of North American Mammals, Birds, Mollusks, Minerals, Tertiary and Cretaceous Fossils, printed on one side for labelling.

The Essex Institute is now printing labels for Corals, consisting of the names of the Orders, Suborders, Families and Genera, after the classification of Professor Verrill, which it will furnish at cost.

The Institute also proposes to print similar labels for the other classes of Radiates and for the Mollusks and Insects.

# PHOTOGRAPHS OF CORALS AND OTHER SPECIMENS.

List of Photographs of Corals, &c., prepared by Prof. A. E. Verrill from original or rare specimens, authentically labeled : —

### STEREOSCOPIC PHOTOGRAPHS.

#### CORALS.

No. 1. *Astrea speciosa* Dana. Original specimen.

No. 2. *Astrea* (Prionastrea) *robusta* Dana. Original specimen.

No. 3. *Favia ordinata* Verrill. *Goniastrea aspera* Verrill. *Fungia papillosa* Verrill. Original specimens.

No. 4 *Cœlastrea tenuis* Verrill. *Pavonia complanata* Verrill. From original specimens.

No. 5. *Madrepora pumila* Verrill. *M. striata* Verrill, *M. prolixa* Verrill. From original specimens.

No. 6. *Prionastrea Chinensis* Verrill. Original specimen.

No. 7. *Distichipora nitida* Verrill and *Stylaster elegans* Verrill. From authentic specimens.

No. 8. *Fungia concinna* Verrill. *F. Haimei* Verrill. From original specimens.

No. 9. *Fungia valida* Verrill. Original specimen.

No. 10. *Allopora Californica* Verrill. Original specimen.

No. 11. *Madrepora efflorescens* Dana. Original specimen.

No. 12. *Pocillopora nobilis* Verrill. Authentic specimen.

No. 13. *Gorgonia Agassizii* Verrill. *G. rigida* Verrill. Authentic specimens.

No. 14. *Madrepora spicifera* Dana. Original specimen.

No. 15. *M. convexa* Dana. Authentic specimen.

#### MISCELLANEOUS.

No. 16. Four Eggs of *Falco anatum*, from Mt. Tom, Mass.

No. 17. *Solaster endeca* and *S. papposus*. Eastport, Maine.

No. 18. *Mastodon* and interior of Prof. J. Wyman's Museum at Cambridge.

No. 19. Three teeth of *Bison*. Quarternary fossils, Gardiner, Me.

No. 20. Nine species of Shells from same formation as No. 19.

### PLAIN PHOTOGRAPHS, LARGER SIZES.

No. 1. *Gorgonia Agassizii* Verrill. Original specimen.

No. 2. *G. aurantiaca* Verrill. Authentic specimen.

No. 3. Four eggs of *Falco anatum*, from Mt. Tom, Mass.

No. 4. Twelve species of moths (*Bombycidæ*) from original specimens described by Dr. A. S. Packard, Jr.

No. 5. Nineteen species *Bombycidæ* from original specimens of Packard.

No. 6. *Samia* (*Platysamia*) *Columbia* Smith, male and female, cocoon and chrysalis from the original specimens.

No. 7. Various rare or new Insects. Collection of A. S. Packard, Jr.

No. 8. Quarternary (Drift) Shells, rare species. Coll. Packard.

No. 9. Drift shells, Maine, Labrador, &c. Coll. Packard.

No. 10. *Mussa crispa* Dana. Original specimen.

No. 11. Section of nest of Common Wasp.

No. 12. Three teeth of *Bison*. Drift fossils, Gardiner, Me.

All the preceding, except Nos. 1, 10 and 11, which are reduced one half, are of natural size.

The stereoscopic sizes will be sent by mail, postage prepaid, at 50 cents each, or the set of 20 for $8. The larger sizes at 75 cents each, or the set of 12 for $8. Address F. W. PUTNAM, Essex Institute, Salem, Mass., or PROF. A. E. VERRILL, Yale College, New Haven, Conn.

## CORRECTIONS TO THE DIRECTORY.

No. 23. E. T. Cox. Should be *North American.*

No. 90. Dr. NEWBURY is now Professor of Geology in Columbia College. For correct address see No. 341.

No. 171. Should be CHRISTIAN FEBIGER.

No. 319. Mr. HYATT, is now residing in Salem. For correct address see No. 482.

## ADDITIONS TO THE DIRECTORY.

### MINERALOGY.

243, a. Prof. G. C. SWALLOW, (State Geologist of Missouri and Kansas), Columbia, Boone Co., Mo.

201, a. J. B. KEVINSKI, Lancaster, Pa. *Local.*

### GEOLOGY.

4, a. WILLIAM ANDREWS, Cumberland, Md. *Local.*

27, a Dr. JOHN DE LASKI, West Falmouth, Me. *Local.* Special, *Glaciers.*

58, a. JAMES HYATT, Bengall, N. Y. *Local.*

### PALÆONTOLOGY.

276, a. WILLIAM ANDREWS, Cumberland, Md. *Local.*

## NOTICES OF PROPOSED WORKS ON NATURAL HISTORY.

TRYON REAKIRT, 335 North 3d street, Philadelphia, Pa.

Is engaged in preparing a Synopsis of the Diurnal Lepidoptera of the Rocky Mountains and Trans-Mississippi Plains. Mr. Reakirt would be happy to receive any specimens from these regions for examination.

ARTHUR M. EDWARDS, 115 John street, New York, N. Y.

Is preparing a work on the Bibliography of the Diatomaceæ; also a List of the described species of Diatomaceæ, with references to the original descriptions and figures; to be published in the Proceedings of the Essex Institute. For further information address Mr. Edwards.

PROF. JAMES HUBBERT, St. Francis College, Richmond, Canada East.

Is preparing a work on the Botany of Canada, entitled A Handbook of the Canadian Flora, being a description of the flowering plants, ferns and mosses indigenous to or naturalized in Canada. A comparison of the Canadian Flora with that of Great Britain, and especially with that of the highlands of Germany will be given, and particular attention will be paid to the limitation of species.

REV. DR. M. A. CURTIS, Hillsborough, N. C.

Is preparing a work on the principal eatable species of Mushrooms and other Fungi, of this country, with colored figures.

# DESIDERATA
### AND
# SPECIMENS FOR SALE AND EXCHANGE.

ALPHEUS HYATT, Essex Institute, Salem, Mass.

Offers to exchange identified specimens of Fresh water Polyzoa, from various localities in New England, for specimens from any other locality.

PROF. H. C. WOOD, JR., Academy of Natural Sciences, Philadelphia, Pa.

Solicits Myriapods or Centipedes, and Phalangidæ or "Daddy-long-legs," from all parts of North America. The former in order to perfect his published monograph on the North American species of the group, and the latter with the intention of monographing them. They should be preserved in alcohol or very strong whiskey in small mouthed bottles. All collections will be returned labeled, if desired. The smaller species are especially desired, also any notes on the habits of any of the species. Prof. Wood has a few copies of his monograph of the Myriapoda of North America which he will give for really valuable collections. The Phalangidæ may be distinguished from the true spiders by the head not being distinct from the abdomen.

DR. S. C. WILLIAMS, Silver Springs, Lancaster Co., Pa.
Minerals and Local Insects for exchange.

H. G. BRUCKHART, Silver Springs, Lancaster Co., Pa.
Local Mollusks and Coleoptera for exchange.

N. VICKARY, 262 Chestnut St., Lynn, Mass. (Taxidermist and Dealer).
Has Birds and other specimens for sale or exchange.

DR. F. STEIN, Museum der k. Universität, Berlin, Prussia.

Has for sale the following Insects from the collection of the late Dr. Schaum. Elatrides (7 boxes), 200 Thalers; Buprestides (4 boxes), 75 Thalers; Chrysomelides (23 boxes), 150 Thalers; Scydmænides and Pselaphides (number not given), 120 Thalers; Cerambycides (European), 40 Thalers.

## THE WALKER PRIZES.

" The following prizes were founded by the late DR. WILLIAM J. WALKER, for the best memoirs, and in the English language, on subjects proposed by a committee appointed by the Council of the BOSTON SOCIETY OF NATURAL HISTORY. The first and second are to be awarded annually; the third, once in five years, beginning 1870.

*First*—For the best memoir presented, a prize of sixty dollars may be awarded. If however, the memoir be one of marked merit, the amount awarded may be increased to one hundred dollars, at the discretion of the committee.

*Second*—For the next best memoir, a prize not exceeding fifty dollars may be awarded at the discretion of the committee: but neither of the above prizes shall be awarded unless the memoirs presented shall be deemed of adequate merit.

*Third*—GRAND HONORARY PRIZE.—The Council of the Society may award the sum of five hundred dollars for such scientific investigation or discovery in natural history as they may think deserving thereof; provided such investigation or discovery shall have first been made known and published in the United States of America;

and shall have been at the time of said award made known and published at least one year. If in consequence of the extraordinary merit of any such investigation or discovery, the Council of the Society should see fit, they may award therefor the the sum of one thou and dollars.

*Subject of the Annual Prize for* 1866-7. "The fertilization of plants by the agency of insects, in reference both to cases where this agency is absolutely necessary, and where it is only accessory;" the investigations to be in preference directed to indigenous plants.

*Subject for* 1867-8. "Adduce and discuss the evidences of the coexistence of man and extinct animals, with the view of determining the limits of his antiquity."

Memoirs offered in competition for the above prizes must be forwarded on or before April first, prepaid and addressed

"*Boston Society of Natural History,*
*for the Committee on the Walker Prizes,*
*Boston, Mass.*"

Each memoir must be accompanied by a sealed envelope enclosing the author's name, and superscribed by a motto corresponding to one borne by the manuscript."
BOSTON, June, 1866.

# THE ENTOMOLOGICAL SOCIETY OF PHILADELPHIA.

This Society, organized in the Spring of 1859, has earned for itself in the short period that has since elapsed, a name and reputation which might be envied by many of the oldest scientific associations in this country or in Europe. Devoted solely to the study of that branch of Zoölogy which its name indicates, its members have infused a spirit of energy and progress into its proceedings, which has accomplished in a few months more than the same number of years have effected in larger bodies endowed with more extensive means. In fact the unavoidable expenses of accumulating and maintaining a collection, of publishing its quarterly journal, &c., have been borne almost entirely by the late Dr. Thomas B. Wilson, of Philadelphia. The generous liberality of this keen student of Nature was checked by his sudden death, after an illness of less than a week, on Wednesday, the 15th of March, 1865. The Society which had been the object of his benevolence had every reason to expect a continuation of his bounty in the form of a bequest, from frequent expressions of his intentions, but his untimely death has placed them in such a position that they cannot continue their former useful career without the aid of all well disposed patrons of science. They are anxious to keep up the prestige of their publications, and for this purpose they ask subscriptions to a Fund of $50 000, of which amount $10 000 has already been realized and funded. To all subscribers of $100 and upwards, an Honorary Membership is tendered, and the publications of the Society will be furnished free of charge during the life-time of the subscriber.

The importance of the work in which this association is engaged, cannot fail to be universally recognized at this time when our crops are yearly destroyed by new and strange insect enemies, whose diminutive size, mysterious transformations, and immense multitude , make us, in our ignorance of their habits, utterly powerless before them. On this subject a monthly bulletin, called the *Practical Entomologist*, is issued by the Society for distribution to any one forwarding their address and *fifty cents* per year to the Secretary of the ENTOMOLOGICAL SOCIETY OF PHILADELPHIA, No. 518 So. 13th street, Philadelphia, Pa. Any one remitting annually a sum of not less than *One dollar*, to the Secretary, E. T. Cresson, will be elected a Contributing Member of the Society, and will receive a Diploma to that effect.

## THE PORTLAND SOCIETY OF NATURAL HISTORY.

We beg to call attention to the following appeal of this most unfortunate society, and to urge the friends of science to give what aid they can as promptly as possible, for every dollar received at this trying time will be more encouraging to the society than larger sums hereafter.

### AN APPEAL TO THE FRIENDS OF SCIENCE.

"For the second time, the Portland Society of Natural History has been visited by a destructive calamity. Its new hall, with the furniture and all its collections, have again been destroyed by fire.

The origin of this Society must be referred to the organization of the Maine Institute in 1836. Though at first struggling with poverty, it was able to secure large collections in Natural History, and a valuable library of scientific books. These were, by the favor of the Government, placed in the Custom House, a spacious hall in that professedly fire-proof structure being assigned to the Society's use. In 1854, the burning of this edifice destroyed every species of property belonging to the Society. Not a vestige of its museum or library was left to serve as a memorial of the past.

But the spirit of the Society was not dead. A few individuals by their persevering labors raised a new cabinet from the ashes of the old. The State granted one half-township of land; subscriptions were set on foot; contributions flowed steadily in; and at last the Society was housed in a noble building, which six months ago it had lifted so far out of debt that it could begin to call this its own. It had a splendid hall of exhibition, fine lecture room and laboratory; while the collections made by some of the most faithful servants of science, or contributed from the East and West, adorned its walls. Publications of high scientific value had been issued from its press. A special Curator had just been regularly engaged, a repair fund gathered, courses of free lectures begun, new members were crowding to its ranks, all the signs of vitality and growth were large, when, in the terrible fire that left, in twelve hours only a desert where the commercial centre of Portland was, everything once more vanished like a dream. The building was isolated, fire-proof apparently, and in the judgment of all safe from harm, until the sweep of that awful tide of flame, which no masonry could withstand, closed in ruin over the cherished results of years of toil.

By this loss, the Society is again stripped of its all. Its insurance proved nearly worthless. Its mortgage debt will absorb the value of the land, the charred ruins and its remaining funds. The library and the fine picture of Humboldt, the splendid gift of the poet Longfellow, alone are saved. In the destitution, which these remnants of former wealth make more painful to consider, the Society is compelled to implore the aid of the friends of science everywhere to enable it to continue its work.

Brethren! whom God has spared the double affliction with which He has visited us, will you grant us your help? Our first need is a home — a building that we may re-consecrate to science. We can repair all other losses better than that. Our city is impoverished — our own selves involved in grief and loss — and if aid does not come from you, we know not where to turn. Will you give us the hand of sympathy— the open hand of benevolence, that we may again have a " local habitation and a name," and go on prosperously in the joyful work of studying and interpreting the book of Nature?

At a meeting of the Portland Society of Natural History, held July 9th, 1866, at the residence of Rev. E. C. Bolles, it was voted that the undersigned be a committee to make a brief statement of facts connected with the history of the Society, and appeal to the friends of science everywhere for aid in this critical condition of its affairs. To this Committee, or to any officer of the Society, all communications upon the subject may be addressed.

OFFICERS OF THE SOCIETY.

WM. WOOD, *President.*

HENRY WILLIS, *Vice President.*

E. C. BOLLES, *Secretary.*

EDW. GOULD, *Treasurer.*

COMMITTEE.

WM. WOOD, M. D.

REV. E. C. BOLLES.

CHAS. B. FULLER.

EDW. S. MORSE.

## THE CHICAGO ACADEMY OF SCIENCES.

The collections of this Academy were partially destroyed by fire on the 7th of June last. The following quotation from the circular issued by the Academy gives its present condition.

"As nearly as can now be ascertained the present condition of the collection and property of the Academy is as follows: —

About half the Mammals and Birds, and nearly all the Skulls, etc., will be saved; the extensive collection of Birds' eggs and nests were entirely destroyed; Fishes and Reptiles are saved; Insects all destroyed with the exception of the Lepidoptera, dried Crustacea and Echinodermata destroyed; Shells and Fossils in great part saved. Very singularly and fortunately, the alcoholic] collection, contained in about 2000 jars, has escaped. The Herbarium, with the exception of the series of the plants of the North Pacific Expedition, is saved. The Library is greatly damaged by water but most of the books will be saved by careful drying and rebinding. The plates of the forthcoming volume of the Transactions, twenty in number, were much injured, and some of the edition may have to be reprinted. The publication of the volume, will not, however, be greatly delayed."

We understand that the Academy will proceed at once to erect a fire-proof building for their collections and library, which, we trust, will, by the aid of kindred societies and friends, soon be larger than before the disaster.

The several disasters to scientific institutions during the last year, should warn all our Societies and Institutions having valuable collections and libraries, to secure fire proof buildings for them at once.*

## THE AMERICAN ASSOCIATION FOR THE ADVANCEMENT OF SCIENCE.

The fifteenth meeting of the American Association was held at Buffalo, N. Y., from August 15th to 21st. Since the commencement of the war these meetings have been suspended and we are glad that they were renewed under such pleasant auspices as attended this meeting, and that the citizens of Buffalo so fully appreciated the value of the Society and were so cordial in their entertainment of its members.

----

· *The collection of the Lyceum of Natural History of New York, was wholly destroyed by the burning of the Academy of Music in May last. We wait for official information in regard to this Society and its future movements.

The officers for this meeting were, *President*, F. A. P. BARNARD, President of Columbia College; *Vice President*, A. A. GOULD, M. D., of Boston; *General Secretary*, Prof. ELIAS LOOMIS, of Yale College; *Permanent Secretary*, Prof. JOSEPH LOVERING, of Harvard College; *Treasurer*, A. L. ELWYN, M. D., of Philadelphia.

The next meeting of the Association will be held at Burlington, Vt., commencing August 21st, 1867. The following are the officers elected for the meeting: *President*, Prof. J. S. NEWBERRY, of New York; *Vice President*, Prof. WOLLCOTT GIBBS, of Cambridge; *Permanent Secretary*, Prof JOSEPH LOVERING, of Cambridge; *General Secretary*, Prof. C. S. LYMAN, of New Haven; *Treasurer*, Dr. A. L. ELWYN, of Philadelphia.

We think that it would, perhaps, have been more advisable to have had the meeting for 1867 held at a more central city, which would have induced a larger number of members from the West and South to attend, but still we hope that, notwithstanding the extreme northern location of the meeting, members from all the states will endeavor to be present and maintain its character as an *American* Association.

## OBITUARY NOTICES.

REV. STILLMAN BARDEN, of Rockport, Mass; well known as a mineralogist, died at his residence on August 7, 1865, of consumption. Mr. Barden was an active and enthusiastic collector and a thorough lover of nature. He had gathered a large cabinet of minerals which will be kept up by his son Edward, who has inherited his father's taste for mineralogy.

DR. SIMEON SHURTLEFF, of Weatogue, Hartford Co., Ct., a general student of nature and especially interested in Botany, Ornithology and Conchology, died at his residence on December 29, 1865.

THOMAS DANIELS, of Cincinnati, Ohio, died in January, 1866. Mr. Daniels was known to many naturalists as a student of Palæontology.

WILLIAM GLEN, of Cambridge, Mass., died of consumption at his home on May 25, 1866. Mr. Glen was a native of Scotland and came to this country in 1854. For several years he was an Assistant in the Museum of Comparative Zoölogy, at Cambridge. He was a person of most remarkable skill in his manipulations, and certainly had no superior in preparing sections and microscopical objects. By the sad death of Mr. Glen science has lost a careful, enthusiastic and faithful worker.

PROFESSOR HENRY DARWIN ROGERS, LL. D., F. R. S., &c., of Glasgow, Scotland, died at his residence in Shawlands, near Glasgow, on Tuesday, May 29, 1866, soon after his return from a visit to the United States. In 1857, Professor Rogers was called to the chair of Regius Professor of Geology and Natural History in the University of Glasgow, which he filled to the time of his death. His intimate connection with the Geology of America, and especially his great work on the Geology of Pennsylvania, will ever endear his name to American Naturalists. He was born in Philadelphia in 1809.

www.ingramcontent.com/pod-product-compliance
Lightning Source LLC
Chambersburg PA
CBHW021808190326
41518CB00007B/502